Ethical Impacts of Using AI for Sustainable Development

Fernando Ortiz-Rodriguez
Tamaulipas Autonomous University, Mexico

Sanju Tiwari
Sharda University, Greater Noida, India

Arturo Mora Olivo
Tamaulipas Autonomous University, Mexico

Sashikala Mishra
Symbiosis Institute of Technology, India

IGI Global
Scientific Publishing
Publishing Tomorrow's Research Today

Vice President of Editorial	Melissa Wagner
Director of Acquisitions	Mikaela Felty
Director of Book Development	Jocelynn Hessler
Production Manager	Mike Brehm
Cover Design	Phillip Shickler

Published in the United States of America by
IGI Global Scientific Publishing
701 East Chocolate Avenue
Hershey, PA, 17033, USA
Tel: 717-533-8845
Fax: 717-533-7115
Website: https://www.igi-global.com E-mail: cust@igi-global.com

Library of Congress Cataloging-in-Publication Data

Names: Ortiz-Rodriguez, Fernando, 1974- editor. | Tiwari, Sanju, 1979- editor. | Mora-Olivo, Arturo, 1963- editor. | Mishra, Sashikala, 1982- editor.
Title: Ethical impacts of using AI for sustainable development / edited by: Fernando Ortiz-Rodriguez, Sanju Tiwari, Arturo Mora-Olivo, Sashikala Mishra.
Other titles: Ethical impacts of using artificial intelligence for sustainable development
Description: Hershey, PA : IGI Global Scientific Publishing, [2025] | Includes bibliographical references and index. | Summary: "Artificial intelligence (AI) is used frequently nowadays, and its impact on different areas is increasing. AI is expected to positively affect sustainability and its effects on achieving sustainable development goals (SDGs)"-- Provided by publisher.
Identifiers: LCCN 2024058262 (print) | LCCN 2024058263 (ebook) | ISBN 9798369386439 (hardcover) | ISBN 9798369386446 (paperback) | ISBN 9798369386453 (ebook)
Subjects: LCSH: Sustainable development--Technological innovations. | Sustainable Development Goals (Project) | Artificial intelligence--Moral and ethical aspects.
Classification: LCC HC79.E5 E74798 2025 (print) | LCC HC79.E5 (ebook) | DDC 338.9/27--dc23/eng/20250320
LC record available at https://lccn.loc.gov/2024058262
LC ebook record available at https://lccn.loc.gov/2024058263
British Cataloguing in Publication Data
A Cataloguing in Publication record for this book is available from the British Library.

Table of Contents

Detailed Table of Contents

This chapter focuses on the themes of innovation & responsibility in AI for sustainable development. It looks at the positive contribution towards sustainability by AI together with the resulting ethical questions. In the chapter, case examples from different fields are discussed with an emphasis and relevance of AI in improving resource efficiency, decreasing the adverse impact on the environment, and promoting socio-economic justice. It also responds to some of the threats that may limit these benefits such as bias, privacy issues, and the digital divide. In this regard, the chapter advocates for a holistic approach to everything to do with the ethical aspect of training to prevent the increase in promulgation of AI gadgets that may hamper sustainable practices. Finally, this chapter hopes to map out a way to the proper use of AI while at the same time pointing out the fact that there is always need for constant watch and adjustments in an attempt to foster sustainable development.

Agriculture development is the process in which the conditions are required for fulfillment of agricultural potential. To achieve these conditions, we need the required knowledge and the availability of technology to allocate the inputs and outputs of the system. Agriculture is the major source of livelihood since most of the laborers who are working in the agriculture sector are shifting their work to other sectors by automating agriculture. It is the key driver of international trade, especially for rice. It also reduces rural poverty by increasing the income of the residents. It not only helps to develop the agricultural sector but also supports the manufacturing sector by providing raw materials. Advancement in agriculture with the help of technology provides an increase in economic growth. The moisture level, fertility, and pesticide level of the soil need to be monitored for precision agriculture. Weather prediction and disease identification are also required for efficient agriculture.

The integration of artificial intelligence (AI) into sustainable development offers transformative opportunities for achieving the United Nations' Sustainable Development Goals (SDGs). AI technologies enhance efficiency in resource management, enable predictive analytics for climate action, and support inclusive economic growth and healthcare access. However, these advancements also introduce significant ethical challenges, including data privacy concerns, algorithmic biases, environmental impact, and accountability issues. This study explores the dual-edged impact of AI in sustainable development, emphasizing the need for ethical frameworks to balance technological innovation with principles of equity, transparency, and sustainability. By examining case studies and existing policy frameworks, this research highlights both the benefits and risks of AI applications in sustainability and proposes actionable recommendations for ensuring responsible and equitable deployment.

Growth and availability of internet with significant higher internet bandwidth and lower cost has fuelled the growth of Over the Top (OTT) platforms. Multiple channels like mobile, tv, set top boxes including gaming platforms need to access the content in secure way. As most of the revenue for an OTT platform comes from subscription and ads, it's crucial to show the right content to authorized users while maintaining the privacy of users following General Data Protection Regulation (GDPR) guidelines. This chapter proposes a security architecture framework for OTT which can be used to achieve higher level of security following zero-trust architecture principles. Each component of the framework from frontend to middleware to backend along with cloud infrastructure is explained in detail with best practices followed for achieving security.

Achieving sustainability and reducing inequities are at the heart of the 17 Sustainable Development Goals set out by the United Nations. When faced with complicated problems, a multidisciplinary team using data-driven methodologies, artificial intelligence, and technology may maximise efficiency, incorporate sustainability considerations, and back up well-informed decisions. This study classifies AI-driven solutions to sustainability problems by doing a literature review, scientometric analysis, and semantic analysis. It emphasises the interface of artificial intelligence and sustainability, as well as important research subjects and regional variations. Based on the results, it's clear that multi-dimensional decision-making requires hybrid methods that incorporate AI, data analytics, and human expertise. Also, the report stresses the significance of making sure AI and big data are used sustainably and the ethical considerations that come with it. To fill in knowledge gaps and propel successful sustainability solutions, collaboration and inclusive research are crucial.

This study presents a systematic analysis of the scientific literature on the relationship between artificial intelligence (AI) and the economy during the period 2020-2024, based on the analysis of 530 documents indexed in Web of Science. Employing a methodology that combines quantitative analyses of scientific production, co-citation networks, and keyword mapping using R software and VOSviewer, the research identifies the main trends, collaboration patterns, and emerging areas in this field. The results demonstrate sustained growth in scientific production, highlighting a significant concentration in three main areas: business digital transformation, supply chain management, and innovation in business models. The fundamental contribution of this study lies in providing an updated and systematic understanding of the state

Preface

The rapid evolution of Artificial Intelligence (AI) has had a transformative effect on various sectors, including those crucial to sustainable development. While AI has the potential to drive significant advancements in fields such as agriculture, energy, healthcare, and urban planning, it also brings forth a host of ethical dilemmas that demand careful consideration. This edited volume, "Ethical Impacts of Using AI for Sustainable Development," explores the intersection of AI technologies and sustainable development, focusing on the ethical implications of their integration into real-world applications.

The chapters in this book reflect diverse perspectives on how AI can be harnessed to achieve the United Nations Sustainable Development Goals (SDGs), while addressing the ethical challenges posed by such technologies. Experts from various disciplines, including computer science, economics, and environmental science, have contributed their insights to provide a comprehensive understanding of AI's role in shaping a sustainable future.

The volume covers a wide range of topics, starting with the balance between innovation and responsibility in AI applications (Chapter 1), followed by specific use cases in sectors such as agriculture (Chapter 2), cryptocurrency (Chapter 5), and manufacturing (Chapter 6). The ethical implications of AI's role in achieving sustainable development are examined in multiple contexts, from zero-trust security frameworks for Over the Top (OTT) platforms (Chapter 4) to the integration of green information systems in the Mexican manufacturing industry (Chapter 6). Further, it delves into emerging technologies like AI-driven solutions in sustainable computing (Chapter 7) and examines the interplay between AI and economic systems through bibliometric analysis (Chapter 8) and its applications in various economic sectors (Chapter 9). Finally, the book concludes with a discussion on how AI can contribute to building sustainable cities and achieving SDG 11 (Chapter 10).

This collection serves not only as an academic resource but also as a call to action for policymakers, industry leaders, and researchers to carefully consider the ethical dimensions of AI deployment. The chapters provide not only a roadmap for the responsible use of AI in sustainable development but also highlight the potential risks of AI's unchecked growth, such as algorithmic biases, privacy concerns, and the exacerbation of existing inequalities.

We hope that this volume encourages further dialogue and research that can guide the responsible integration of AI into sustainable development strategies, ensuring that technological progress aligns with the broader goals of social equity, environmental stewardship, and economic justice.

CHAPTER OVERVIEW

- Chapter 1: Balancing Innovation and Responsibility: AI in Sustainable Development

This chapter examines the relationship between innovation and responsibility in AI's application for sustainable development, exploring both its benefits and the ethical issues it raises, such as biases, privacy concerns, and the digital divide. Case studies are presented to highlight AI's potential and limitations in contributing to sustainability.

- **Chapter 2: Intelligent Agriculture**

Focusing on the role of AI in revolutionizing agriculture, this chapter discusses how AI-driven solutions improve efficiency, productivity, and sustainability in farming practices. It highlights the impact of technology on rural livelihoods and examines the challenges of integrating AI in agricultural systems.

- **Chapter 3: Ethical Impacts of Using Artificial Intelligence in Sustainable Development Challenges and Opportunities**

This chapter explores the dual-edged impact of AI on sustainable development, highlighting its potential to achieve the SDGs while also posing significant ethical challenges. It discusses the need for ethical frameworks to balance AI's benefits with its risks, using case studies and policy frameworks as examples.

- **Chapter 4: Zero Trust Architecture based Enhanced Security Framework for OTT in Hybrid Cloud Environment**

The focus of this chapter is on ensuring security in Over the Top (OTT) platforms by employing zero-trust architecture. It explores the ethical implications of data privacy and security in hybrid cloud environments and outlines best practices for implementing secure AI-driven systems.

- **Chapter 5: Cryptocurrency Price Prediction: A Comprehensive Analysis of Social Media Sentiment Analysis**

By employing advanced AI models, this chapter explores how sentiment analysis from social media platforms, like Twitter, can be used to predict cryptocurrency market movements, focusing on Bitcoin. The ethical concerns of data manipulation and market prediction are discussed.

- **Chapter 6: Integrating Green Information Systems as Organizational Performance Tools for the Mexican Manufacturing Industry**

This chapter examines how Green Information Systems (GIS) are integrated into Mexican manufacturing firms to drive sustainability. It looks at how GIS technologies enhance environmental strategies and contribute to economic performance, providing empirical evidence on their effectiveness.

- **Chapter 7: Exploring the Convergence of Artificial Intelligence and Sustainable Computing**

Here, AI's convergence with sustainable computing is explored, discussing how data-driven decision-making, AI, and technology can contribute to solving sustainability challenges. Ethical considerations surrounding the use of AI in sustainability solutions are also addressed.

- **Chapter 8: Bibliometrics of Artificial Intelligence and Economy**

This chapter offers a systematic review of the scientific literature on AI's impact on the economy, identifying key trends, research areas, and collaborations in the field. It provides a comprehensive understanding of how AI is transforming economic models and decision-making processes.

- **Chapter 9: The Use of AI Applications in the Economy**

Focusing on AI's applications in various sectors of the economy, this chapter explores how AI tools are used to improve efficiency in finance, marketing, and decision-making processes. It highlights the ethical implications of these technologies, including their impact on workforce dynamics and consumer privacy.

- **Chapter 10: Sustainable Cities: A Pathway to Achieving UN Sustainable Development Goal 11**

The final chapter explores how AI and smart city technologies can help achieve SDG 11 by making cities more inclusive, safe, resilient, and sustainable. It discusses urban sustainability challenges and the role of technology in addressing them, highlighting AI's contribution to sustainable urban development.

Chapter 1
Balancing Innovation and Responsibility:
AI in Sustainable Development

Roop Kamal
https://orcid.org/0009-0001-8334-0101
Chandigarh University, RIMT University, India

Shipra
RIMT University, India

Yashmin Sofat
A.S. College, Khanna, India

ABSTRACT

This chapter focuses on the themes of innovation & responsibility in AI for sustainable development. It looks at the positive contribution towards sustainability by AI together with the resulting ethical questions. In the chapter, case examples from different fields are discussed with an emphasis and relevance of AI in improving resource efficiency, decreasing the adverse impact on the environment, and promoting socio-economic justice. It also responds to some of the threats that may limit these benefits such as bias, privacy issues, and the digital divide. In this regard, the chapter advocates for a holistic approach to everything to do with the ethical aspect of training to prevent the increase in promulgation of AI gadgets that may hamper sustainable practices. Finally, this chapter hopes to map out a way to the proper use of AI while at the same time pointing out the fact that there is always need for constant watch and adjustments in an attempt to foster sustainable development.

DOI: 10.4018/979-8-3693-8643-9.ch001

INTRODUCTION

Figure 1. Sustainable Innovation

Sustainability innovation has emerged as a significant concern for current organizations in order to address concerns that include environmental accountability and customer and regulatory demands (Steinhoff, 2024). It refers to the incorporation of sustainability concepts like the reduction of the adverse effect on the environment, promotion of the company's and society's welfare, and the sustenance of the financial health of the business within the key innovation processes of a firm (Cooper, 2024). This transformation not only must help to mitigate a host of global threats ranging from climate change to resource scarcities but also can provide established corporate players with fresh sources of value creation, competitive advantage, and sustainable organizational growth (Abbas et al., 2024).

One of the main reasons is the shift in consumers' attitudes as they are now focused on environmental and social issues of products (Singh & Maheswaran, 2024). An increasing pool of informed consumers is putting pressure on organisations to declare their manufacturing, sourcing, and emission-management policies (Kamal, 2024). This has resulted into what is known as green marketing where firms market their ecofriendly status as part of what they offer to the market (Glaeser & Lang, 2024). Furthermore, the governments of various countries continue to set higher standards and implementing overall rules and restrictions on carbon emissions, control over the wastes and the usage of resources meaning that the companies have to find ways of sustainability innovation or they face damaging penalties and reputational losses (Plooy, 2024).

Innovation has to be sustainable, and this is only possible in most cases is the change of the existing business models. The key sustainable business model that involves reducing waste, renewing the usage time of a product and remanufacturing of products may create new revenue streams for a firm or organization with a minimal negative impact on the environment (Salaheldeen & Battour, 2024). One good example is the change in ownership from possessing the actual product to availing services such as for a fee of subscription or hire-purchase agreements. This approach makes sure that the Businesses focus especially on crafting items that are long-lasting, can be fixed easily and even upgraded and so, it also enhances sustainability (Hassan et al., 2024).

The second way that sustainability is integrated into the innovation process also encourages the development of technologies including renewable energy, green manufacturing, or sustainable material. For example, energy firms are spending billions of dollars on renewable energy sources such as wind, solar and hydro power so as to minimize the impacts of climate change that are brought about by the use of fossil energy (Sawyer & Henriksen, 2024). Similarly, as for product design, manufacturers are looking to use bio or recycled material in the products that they build to reduce environmental impact of production lines. AI and IoT are two digital technologies that can help businesses go further in resource management, supply chain traceability and the elimination of inefficiencies (Kamal, 2024).

In addition, there is one more point that sustainable innovation is not only in products and technologies it is also in organizational culture and leadership. Building sustainability and innovation into organizational structures and management from the Boardroom to the shop floor leads the organisation to be more responsive to global challenges (Sengupta et al., 2024). Some executives who dedicate their efforts to create superior value over tangible monetary gains can establish long-lasting stakeholder relationships and trust from the customers, shareholders and other personnel. For this reason, such businesses stand higher chances of succeeding in a sophisticated environment that has turned a new page on going green (Hassan et al., 2024).

If organizations want to thrive and remain competitive, and especially with the Global Manager's constant changes the world become, they cannot afford not to practice sustainable innovation (Steinhoff, 2024). Thus, starting from environmental and social perspectives, companies can actually solve the key problems of global sustainability but also generate new value and growth; moreover, they will be able to build a competitive advantage. The concept of sustainable innovation helps to consider the idea of the future business model, which is based not only on profit-orientation but also on the sustainable approach to the company's action (Cooper, 2024).

INNOVATION AND SUSTAINABLE DEVELOPMENT

Innovation is central to sustainable development that seeks to find better ways of addressing the social, environmental and economic problems that exist in the world today. Sustainable development according to the United Nations as development that is required to fulfill the needs of the current population without endangering the ability of the future generations to fulfill their want, is the concept that has forced many organizations and societies to develop new thinking's to support economic growth and progress in social setting that is friendly to the environment (Abbas et al., 2024).. Within this context, innovation is not restricted to the development of new technologies, of new products, but to a completely different conception of growth, of consumption, and use of resources so as to enable a sustainable development (Singh & Maheswaran, 2024).

There is one of the essential pillars that are being advanced through innovation towards sustainability that is transitioning to a low-carbon economy. The development of solar, wind and hydroelectricity technologies are getting enhanced and cost optimal which will reduce dependency on conventional sources of energy (Parolin et al., 2024). This is important also to check on Green House emissions, fight climate change and maintain energy security. Moreover, the progress in the battery technologies, sophisticated networks, and electric cars contributing toward improving the interconnection of renewable energy into routine use, making a possibility of sustainable energy viable for both the commercial as well as the residential segments (Hassan et al., 2024).

Another case of how innovation enhances sustainable development is the case with circular economy. It also contrasts with the more familiar linear economic model that involves resources extraction, consumption, and disposal, by trying to maintain materials in use for as long as possible and by so doing designing for durability, reuse and recycling (Nahar, 2024). Changes in product design, the use of new materials and advanced waste management technologies are essential prerequisites for this process (Singh & Maheswaran, 2024). Many businesses that have implemented circular economy strategies have two benefits by decreasing their impact on the environment while at the same time gaining new opportunities for business and markets for the repurposed materials (Glaeser & Lang, 2024).

Another concept to be reviewed as far as sustainable development is concerned is social innovation which relates to the social aspects that are concerned with sustainability. It includes defining new approaches for addressing social wants and necessities along with the enhancement of living standards and the eradication of disparity (Nahar, 2024). For example, microfinance innovations have offered financial services primarily to the deprived customers with the primary aim of helping individuals as well as communities to participate in economic activities and thereby enhance their standard of living. In the same manner, education, health and housing innovations play central role in curtailing social injustice and making development work for all people (Iftikhar et al., 2024).

Sustainable agriculture is another important area where the focus is made on innovations. Food demand rises as the population expands globally hence the need for change or improvement in the manner in which Agricultural technology is done so as to avoid consequences such as deforestation, water shortages, and degrading of soils (Singhet al., 2024). Through precision farming, biotechnology and agroecology experts are enhancing resource efficiency, productivity and diminishing the use of injurious chemicals in the production of food crops hence the food security and environmental conservation (Kamal, 2024).

Furthermore, computer technologies such as AI, blockchain, and the IoT continue to finds application in sustainable development difficulties. They enhance the collection and analyses of data, in the use of resources like water, energy and land in the farming practices. It also assists in enhancing transparency and accountability towards the suppliers, and all the users to make the right decisions towards sustainability goals (Nahar, 2024).

If it is true that all human beings, communities and countries of the world are faced with challenges of development, then innovation is a vital factor in achieving sustainable development, as it offers solutions in the form of tools and approaches. From technology, new forms of business models and socially innovative solutions, there must be integration of sustainability into the innovation processes in order to foster an equitable, resilient and sustainable future (Sawyer & Henriksen, 2024). It is a future where growth is sustainable and improving value for people, next to profitability, the essence of global growth and a focus on developing innovation can deliver (Shah et al., 2024).

AI AND SUSTAINABLE DEVELOPMENT

Figure 2. Sustainable Development with AI

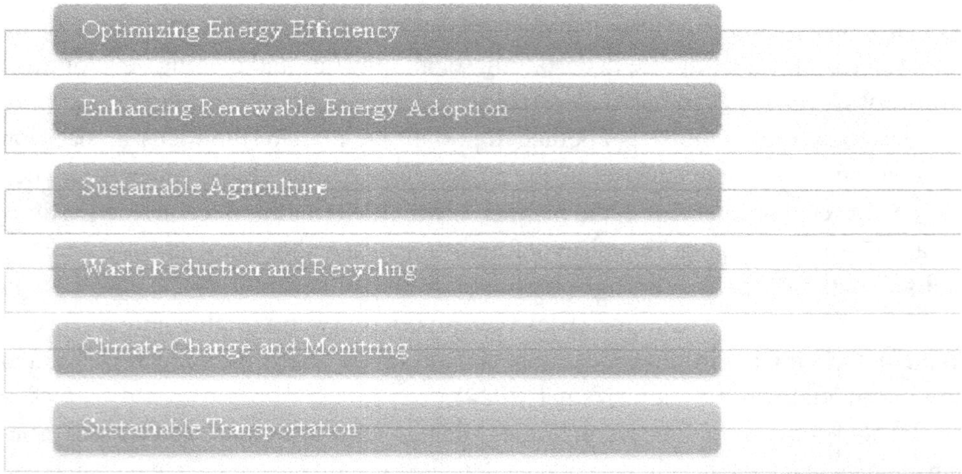

AI is expected to deliver rapid advancements towards sustainability because technology can come up with unique solutions to complicated environmental, social and economic issues of the world (Plooy, 2024). With the use of applied Artificial Intelligence solutions in business, government and other organizations, useful resources can be applied, and consumption can be minimized, while making decisions that reflect sustainable practices in the society (Nahar, 2024).

Climate change is one of the most effective areas that can be applied based on artificial intelligence for sustainability (Lodhi et al., 2024). The application of AI technologies is effective for the purpose of measuring and controlling carbon emissions employing predictive analytics, smart grid management, optimum energy efficiency, and other factors (Singhet al., 2024). For example, in AI in Energy Management Systems, real-time data are employed to forecast energy demand to allocate energy in a way that utilizes renewable energy source like Wind and solar in the most efficient manner. Automating heating, cooling and lighting in the building also becomes the function of AI and this results in lowering energy use, without impacting the environment or performance of occupants (Abbas et al., 2024)..

AI is also revolutionizing the transport industry, a key driver of global greenhouse gas emissions. Self-driving cars, mainly driven by artificial intelligence, have the capability of solving the problem of traffic jams and poor efficiency due to unnecessary time wastage in circulation around the city or extra time wastage in traffic holdup (Parolin et al., 2024). AI is also significant to the creation of electric automobiles and its batteries as well as charging networks. Further, with the use of AI logistics systems supply chains can be made efficient resulting in minimum environmental cost in terms of transportation and ineffective transportation of empty loads (Iftikhar et al., 2024).

Food production is another sector where AI is being applied with the objective of enriching agricultural production and at the same time reducing the adverse effects on the environment. Precision agriculture where farmers use artificial intelligence tools in sensors and data help farmers to monitor the health of the soil, the weather, and the growth of crops (Lodhi et al., 2024). It contributes towards rational utilisation of water, fertilizers and pesticides with the resultant effect of minimising pollution. AI can also help to predict diseases and pests attacks on crops, so that farmers can apply protective measures that will help them avoid using toxic chemicals. As will be elaborated later, these innovations promote sustainable agriculture since they enhance production while minimizing on the use of resources from nature (Sawyer & Henriksen, 2024).

AI too has a significant use in the achievement of circular economy concepts and goals of reducing wastage and maximizing the utilisation of products and materials (Kamal, 2024). AI technologies can enhance the process of sorting and recycling of waste materials which facilitates the separation of valuable material from waste streams. Also, AI-based applications prompt corporations to create products that are longer-lasting, cheaper to recycle, or easier to disassemble as a way to decrease the negative effects of manufacturing and their use (Cooper, 2024, Steinhoff, 2024).

When it comes to the protection of biological and landscape diversity AI allows for observing the changes and studying the dynamics of the ecosystems and the species, the detection of threats including deforestation, poaching and destruction of habitats (Glaeser & Lang, 2024). For example, satellite imagery analysis using artificial intelligence techniques could be used in identifying the changes in land use and forest cover, so as to improve the conservation practices. Likewise, AI can process audio and video data coming from sensors installed in polar regions to monitor A-listed species and(or) poachers in real-time (Sengupta et al., 2024).

Furthermore, it helps sustainable development as indicated by the concept of smart cities through use of AI. Here are some cities that employ AI to optimize the city's system, entering the city's system such as traffic, energy and waste efficiently while reducing the use of resources as well as the negative effects on the environment. For instance, it can be used in traffic with the aim of signal control or managing of water and waste that is aimed at the decrease of traffic jam and emission and avoiding of waste (Lodhi et al., 2024).

However, there are issues that government and business stakeholders applying AI must address and overcome to guarantee that the use of AI is sustainable. Machine learning and deep learning AI models need significant computational resources to build and implement, mainly because they are complex. This concern is going to be dealt with innovations in AI design and consequently improved efficiency of algorithms and hardware (Plooy, 2024). Also, the questions of ethical nature arise as it is necessary to apply AI in a correct way not causing negative impacts on society, for example, in terms of widening the social gap or worsening the conditions for the planet.

AI can therefore open up a vast avenue to support sustainability as it holds the capability to manage resources, minimize effects on the environment while improving decision making on companies' aspects across numerous industries (Steinhoff, 2024). Society has seen artificial intelligence making its way into almost all sectors, including energy and transport, agriculture and conservation to mention but a few. However, the green and ethical impact of AI themselves cannot be ignored so the sustainable usage of AI in the long-run can be achieved (Felicetti et al., 2024). Thus, when applied properly, AI has opportunities opening a greater future not only for humanity but for the entire planet as well.

CASES

1. Energy Management and Efficiency

- Google's Data Center Efficiency: Google is one of the firms that has incorporated AI to help cut energy use in data centers which are known to be quite energy-demanding (Singhet al., 2024). In this case, DeepMind training AI algorithms helped Google to regulate cooling systems in real-time, and the component of energy consumption for it decreased to 40%. In addition, the firm's carbon dioxide emissions were also cut especially concerning operation cost that was trimmed down by this innovation (Shah et al., 2024).
- Smart Grids: Smart grids use AI to address the problem of supply and demand of energy. Machines that use artificial intelligence can review those tendencies and distribute electricity from renewable resources such as the sun and wind. This also helps in reducing the use of fossil-based energy sources, increases energy intensity, and decreases the overall emission of greenhouse gases (Sengupta et al., 2024). For instance, Iberdrola, which is the largest electric company globally, has implemented the application of AI in the energy sector where it aims at calculating various energy requests and generates the best means to adopt renewable energy to the grids.

2. Agriculture and Food Production

- Precision Agriculture: These are some technologies; John Deere AI tractors, and IBM's Watson Decision Agriculture Platforms help farmers to monitor and analyze the soil condition, weather conditions, and crop conditions in real-time. This results in effective utilisation of the products that are used such as water and fertilizers, hence reducing the impacts on the environment as well as improving the yields. Through proper provision of the right amounts at the correct time, the systems aforementioned necessitate the moderation of water use, minimize the leakage of chemicals and improve the effectiveness of the soil hence the sustainability (Parolin et al., 2024, Singhet al., 2024).
- Crop Disease Detection: With the use of AI algorithms, crop diseases are being identified and pest attacks predicted early so that farmers can act accordingly. For instance, PlantVillage uses Artificial Intelligence to diagnose diseases on crops by analyzing images in order to assist farmers, particularly in Africa, to adopt a more efficient method in farming that would help in avoiding crop loss (Steinhoff, 2024). This enhances access to food and at the same time the farming of food crops is done sustainably (Hassan et al., 2024).

3. Waste Management and Circular Economy

- Automated Recycling: Automatic sorting robots, which have been designed by Californian AI company AMP Robotics, employ AI to recognize and sort recyclable materials in a more effective manner as compared to human workers (Simeon et al., 2024). This leads to the reduction of contamination in recycling streams and thus improves on the overall efficiency of the recovery rate of sizes of valuable materials. Due to enhanced efficiency in recycling, such systems minimize the quantity of waste that is dumped in the dump sites hence decreasing the effects of waste on the environment (Felicetti et al., 2024).
- Circular Economy Innovations: Google, for instance, and HP are among firms that are employing AI in designing for durability, ease of repair, as well as ease of disassembly as a way of promoting the concept of the circular economy. Automated algorithms assist in the designing of the processes to reduce resource consumption and make products that are easy to dematerialize and reclaim. This also helps in the reduction of waste, use of recycled or recycled materials in the production hence reduce the overall carbon footprint (Kamal, 2024).

4. Climate and Environmental Monitoring

- AI for Forest Conservation: In the context of deforestation in the Amazon Basin, AI is being applied to identify the occurrence of deforestation and track violations of loggers. Such nonprofit organizations as the Rainforest Connection employ the use of AI algorithms to mine audio data from the sensors in the forest. The system is able to analyze the noises produced by chainsaws and transmit real-time information to authorities regarding the activity thus eliminating cases of unauthorized logging and reduction of the destruction of some species (Lodhi et al., 2024).
- Wildlife Monitoring and Conservation: Some of the positive impacts of AI include in identification of endangered species and, therefore, efficient management of ecosystems. For instance, there is Wildlife Insights which is an application that employs the use of AI in identification of animals from millions of camera trap pictures in real time. It also assists researchers to monitor animals, changes in the ecosystem and then take measures to conserve animals on the verge of extinction (Glaeser & Lang, 2024).

5. Water Management

- Smart Irrigation Systems: Such systems as those used in precision irrigation systems by Netafim involve use of sensors that collect data on moisture content of the soil and possibly the requirements of crops. These systems also enable targeted delivery of water whereby the water is only used where there is need hence conserving the scarce resource of water. On this front, this technology has proved very useful, especially in regions that experience frequent droughts, because it enables the farmers to continue with their farming practices, but use little water (Saxena et al., 2024).
- Urban Water Management: For instance, some urban areas such as Cape Town in South Africa have employed the use of artificial intelligence to enhance its capability of managing water resources well especially during dry seasons. AI systems take data from sensors installed in the water supply and demand systems to estimate usage and identify loss. By increasing water distribution efficiency and finding areas which can be improved, AI ensures that the available water is rationed fairly (Iftikhar et al., 2024).

6. Transportation and Smart Cities

- Autonomous Electric Vehicles: Self-driving cars also known as AI-driven AVs and the use of electric cars, or EVs, is minimizing fuel usage and emissions. AI is currently being used in AVs by companies like Tesla and Waymo and these are believed to have the capability of predicting the best routes, avoiding traffic jams and hence reducing emitters (Simeon et al., 2024). Conventional Electric AV that is operated by renewable resources is the future of urban transportation (Kaur et al., 2024).

- Smart Traffic Management: Artificial Intelligence has found application in controlling traffic and avoiding congestion in the urban areas. For instance, the Surtrac traffic management system in operation in Pittsburgh applies Artificial Intelligence in determining and adjusting traffic lights. This also removes time wastage, reduces exhaust gas emissions and enhances the organization of transport systems in congested cities (Lodhi et al., 2024).

A HOLISTIC APPROACH TO THE ETHICAL ASPECT OF AI TRAINING FOR SUSTAINABLE PRACTICES

Figure 3. Ethics and AI in sustainable Development

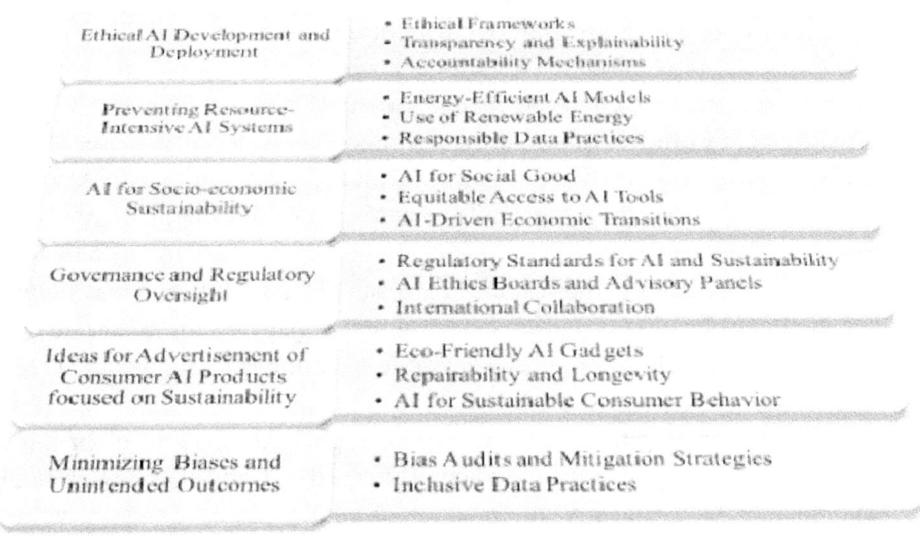

Given its growing integration into the fabric of our society, it is now important to understand how to accompany the emergence of AI with principles that will make its development, deployment, and utilization more sustainable. This implies that there should be a problem-solving approach whose solution is not a single but rather an integrated method of handling AI since it is capable of posing numerous hazards and has implications on the society (Abbas et al., 2024). The increase in daily accessibility of AI devices and setting up of systems may bring abundant benefits in future but it can act a barrier to sustainability if not controlled by proper ethical norms (Felicetti et al., 2024). Here's a holistic approach to ensuring that AI contributes positively to sustainability. Here's a holistic approach to ensuring that AI contributes positively to sustainability:

1. Ethical AI Development and Deployment

Thus, it can be evaluated that the main issues in the context of implementation of AI systems are the key sustainability and ethics considerations, which need to be predicted and included in the development and deployment stages of the AI systems (Cooper, 2024). This entails promoting Ethics in design and Construction, being transparent and again, being accountable all the way from the construction phase to the disposal phase of an AI.

- **Ethical Frameworks:** It is imperative that organizations and governments use appropriate ethical AI guidelines that take into account sustainability, efficiency of resource use and longevity of the common good. Ethical regulation should step beyond the traditional business values and explore how the AI system affects the environment and society (Kaur et al., 2024).
- **Transparency and Explainability**: AI models, especially those adopted with massive amounts of data, need to be incorporated into transparency and interpretability into the systems. Transparency enables the processes which lead an AI to make specific decisions and recommendations to be audited for their sustainability implications. For example, an AI system that suggests the energy utilization or resources consumption patterns must have an easy-to-understand explanation for the environmental impact optimization (Plooy, 2024).
- **Accountability Mechanisms**: Creators of applications and developers must be made Fuller responsible for the actual effects they bring on the environment and society. Consequently, the practice of audits, impact assessments, and reporting can help to guarantee that AI will remain free of ethical concerns and does not amplify unsustainable behaviors (Kamal, 2024). Furthermore, ethical AI projects must contain ways of redressing for unfavorable results or mitigate the effects of unanticipated impacts online in real-time.

2. Preventing Resource-Intensive AI Systems

Training large and more complex AI models, especially through the use of machine learning and deep learning techniques, requires huge power and water to support the process (Shah et al., 2024). The implementation of the solution has a strong negative associative response with environmental impact and climate change mitigation, threats to which have become imminent.

- **Energy-Efficient AI Models**: There should be a push factor to see to it that AI models which are being developed should be energy friendly. Data analysts and developers can focus on additional refinements on algorithms that will make computations easier hence reducing the energy needed in the training process (Iftikhar et al., 2024). It is revealed that existing techniques such as federated learning, and transfer learning or model pruning can be effective in reducing the levels of AI emissions without having to compromise with its efficiency (Kaur et al., 2024).
- **Use of Renewable Energy**: To reduce AI training and operating costs, there is a need to capture renewable energy where feasible during the deployment of these systems. Some recommendations include large AI-oriented firms using extensive data centers or researching AI, using green energy or transferring to green energy as soon as possible and avoiding environmental pollution. For instance, Google made a commitment to power 100% of its data centers with carbon-free energy by 2030, a good example of how technology firms are modeling proper AI growth (Hassan et al., 2024).
- **Responsible Data Practices**: AI systems are fed by huge data sets, and the process of data gathering, storage, and proper handling takes a lot of effort. The general environmental impact of artificial intelligence can considerably be minimized through ethical data practices such as avoiding the collection of irrelevant data and the management of data through environmentally sound data centers (Kamal et al., 2024).

3. Socio-Economic Sustainability

The development of AI innovation to support the previously mentioned concept of socio-economic sustainability for healthier and more efficient society thus put into use the concepts of inclusive and equitable AI (Alhosani & Alhashmi, 2024).

AI should therefore be developed beyond the environmental impacts to socio-economic Responsible AI. This involves making sure that AI doesn't make things worse by reinforcing disparities or by offering gains that are largely accumulated by the affluent and potent while people at the base of the social hierarchy suffer (Kamal, 2024).

- **AI for Social Good**: AI must be used in matters concerning social & environmental challenges e.g., poverty, climate change, and education & healthcare needs. Therefore, when companies and governments are directing AI innovation towards solving global challenges, they are always guaranteed that AI is going to contribute towards socio-economic justice (Salaheldeen & Battour, 2024). For instance, AI can be applied to solve such issues as access to clean water, efficient distribution of renewable energy and agriculture technologies in areas where they are needed.
- **Equitable Access to AI Tools**: Its democratization means that its potential benefits from AI technologies should be distributed among all the members of society. This means making certain that developing nations and the vulnerable groups are not locked out out of technologies such as the use of Artificial Intelligence for sustainability and even upliftment of their standards of living (Simeon et al., 2024). Educating the public, training the government authorities, and providing the necessary infrastructure in the areas of the developing world should be the priorities for governments and international organizations, which desire to get the maximum effective contribution from the AI facilitation (Felicetti et al., 2024).
- **AI-Driven Economic Transitions**: While computerization is already a growing reality in industries such as manufacturing, logistics, and agriculture, there is a risk that the automation of jobs might lead to the displacement of the workers and affect the local economies. Ideologies of ethical artificial intelligence should contain processes by which people who might be replaced by machines can be retrained. This may assist in preventing the further marginalization of the poor and help ensure that the Fourth Industrial Revolution's advent through AI serves the economic justice to common folks and supports their livelihoods in the long haul (Iftikhar et al., 2024).

4. Governance and Regulatory Oversight

Indeed, Havenstein and others argue that powerful institutions and rules need to be in place to guarantee the application of innovative technologies, such as AI, align with the principles of sustainable development and general ethical norms.

- **Regulatory Standards for AI and Sustainability**: There is a need for governments to set up definite policies that require sustainability concerns in the development, education, and implementation of all AI systems. This could comprise environmental policies for data centers, constraints on use of energy in AI training procedures and codes on ethical reviews of AI projects (Hassan et al., 2024).

- **AI Ethics Boards and Advisory Panels**: With the introduction of AI systems by various companies, it is recommended that the companies establish AI ethic boards or AI ethic advisory committees that comprise of ethicists and other professionals from other fields ranging from environmental science, social justice, and ethics (Kamal, 2024). These boards can offer recommendations to the effects of AI projects to the environment and to the people so as to consider the ethical issues at an early stage.
- **International Collaboration**: Sustainability challenges, for instance, climate change are the problems that have no borders and can only be solved through the collaboration of countries and the world at large. It is suggested that governments, companies and non-governmental organizations should join to set global Aethics/frameworks that would raise sustainability and fairness to the uppermost degree (Shah et al., 2024). This could avoid the solving and exportation of unsustainable AI technological solutions across the world and make sure that AI works to the realization of sustainable goals.

5. Ideas for Advertisement of Consumer AI Products focused on Sustainability

AI-enabled products aimed at consumers including smart home gadgets, wearable devices and personal assistants are increasingly becoming common (Kamal et al., 2024). What may be convenient, and allows for constant connectivity, has a constant environmental price tag associated with extraction of natural resources and supplies, manufacturing and energy usage.

- **Eco-Friendly AI Gadgets**: It will be strategic for the industries to invest in AI gadgets which are eco-friendly in their design. This is as follows; the use of environmentally friendly packaging by using either recyclable or biodegradable materials, adopting low energy devices, and adopting designs that enhance product durability. There are firms such as Fairphone that have opened up a niche in sustainable electronics to show that one can make complex gadgets that are also green (Alhosani & Alhashmi, 2024)..
- **Reparability and Longevity**: The consumer electronics industry is characterized by cyclic production and sales of products with short useful lives with design for obsolescence. There is a principle of ethical AI development that has to do with designing gadgets that can be repaired and upgraded, to have longer life spans (Felicetti et al., 2024). The idea of incorporating artificial in-

telligence into the devices should not in any way make them become outdated after several years of usage but instead enable them to have compatibility with other advanced technologies.

- **AI for Sustainable Consumer Behavior**: Using AI-powered gadgets one can also encourage sustainable consumption among the population. For instance, smart thermostats which include Google Nest or energy management systems such as Sense assist the consumer to control their energy usage at home. In other circumstances, AI can also help in the efficient use of water, the reduction of food spoilage or the promotion of environmentally friendly means of transportation, which positively impacts sustainability (Glaeser & Lang, 2024).

6. Minimizing Biases and Unintended Outcomes

Another moral issue concerning the creation of AI is bias, which might cause AI's in equality that goes against socio economic justice.

- **Bias Audits and Mitigation Strategies**: AI systems have to be trained using a diverse dataset lest it end up discriminating against specific groups of people. It is, therefore, necessary for organisational bias analysis to be routinely run for an AI system to check for any emerging form of an unequal bias that can impact the allocation of resources, environment, or business opportunities (Kamal et al., 2024).
- **Inclusive Data Practices**: Ethical AI development should take into consideration the matter of sample, cases collecting: it should embrace all the people, including those who are discriminated in society (Salaheldeen & Battour, 2024). This can assist in avoiding entrenchment of the pre-existing disparities when it comes to the implementation of the design of the AI systems and make sure that even in its utilization, it enhances fairness in society.

BALANCING INNOVATION WITH CONSTANT VIGILANCE

Since AI has found its way into different aspects of life, it requires a proper and sustainable solution towards its advancement. Meaning that for AI to anchor itself in a positive manner into sustainable development, there is the need to first define the framework to ethical AI. This framework should be centered on sustainability impacts, social concerns and long-term organizational effects. AI solutions must therefore be developed in a way that does not contribute to a negative impact on the environment, for instance by using a lot of power when developing different models.

Besides this, the application of the solutions should deal with industries such as agriculture, transport and manufacturing that impact the societies' environment negatively (Glaeser & Lang, 2024). AI energy consumers such as model compression to reduce computational resources or federated learning that helps reduce data transmission can go a long way in reducing AI's impact on climate change, in addition to moving data centers and IT structures to renewable power sources. However, AI has to be made more responsible, especially working towards socio-economic justice to make sure it helps in the elimination of inequalities as opposed to increasing them. This can be done through the application of AI in enhancing the supply of basic public utilities including health, education and economic opportunities for the disadvantaged areas to be made available to reduce isolation and lack of equality (Alhosani & Alhashmi, 2024). The other equally critical factor is the control and regulation of artificial intelligence systems. International communities and governments must work together to advance effective policies that would ensure that advancements in artificial intelligence are made in line with the intensions of sustainable development. This includes forming autonomous AI ethics councils to give continuous monitoring of the initiatives involving AI and give directions concerning the ethical, societal and environmental impacts of AI. It is recommended that the standards of sustainability be checked from time to time in case of non-adherence and necessary changes be made to AI. A key aspect that is needed for ensuring accountability is transparency and explainability in the AI systems selected. From the energy management case, users have to know how these AI systems come to their decisions so that AI does not encourage or lead to unsustainable processes. Ethical use of AI is not limited only to B2B use but likewise has to be incorporated in C2C use of AI products (Kamal, 2024). Smart devices ranging from home consumer products to wearable technology need to be designed to be sustainable by justifying the use of raw materials such as metals and plastics which can be recycled, reducing power consumption, and making products easy to repair to further reduce e-waste consumption (Hassan et al., 2024). To address these concerns as presented, they would call for a systems approach where the implementation of interventions is never ending given the dynamic nature of technology (Cooper, 2024). AI systems have to be flexible and capable to respond to new issues and emerging technologies as well as new sustainability requirements. This is something that needs to be done constantly to ensure that there are no adverse effects that come with the increased use of artificial intelligence. AI technologies are dynamic, and hence they should undergo checks from time to time to guarantee that they will in support of sustainability instead of pressing on the gains which are without any doubt are corrosive to the environment and society (Shah et al., 2024). In this manner, constant supervision of the AI and continuous modification of the AI technology shall ensure that the AI technology integration in the SD context does not result in the deterioration of the environment, resources and the society (Kamal et al., 2024).

CONCLUSION

The chapter shows that the opportunities of the AI's utilization in the context of sustainable development should be balanced with the responsibility for their application. It is important to note that, with all the positive effects that can come from the use of AI in identifying the solutions to these big issues around the world, ranging from resource consumption to carbon footprints, to social justice, that if not well managed, AI has the potential to be dangerous (Simeon et al., 2024). To ensure that the AI-enhanced systems work effectively and for the greatest benefit, it is important to underline the specific principles of ethicality where the environmental considerations, energy efficiency and social inclusiveness were considered as a core set of propositions during the AI development processes. There is a need to design AI systems and algorithms with the aim of reducing the effects that these systems have on the natural environment, use efficient energy consuming algorithms in the AI systems as well as incorporating the renewable energy sources in the structures of AI systems.

At the same time, it is necessary to turn one's attention to the social consequences of Artificial Intelligence. AI has to bring fairness and make people with opportunities for personal advancement and improvements in their lives, including access to knowledge and healthcare for the underprivileged. This calls for constant supervision so that governments, organizations, and other independent bodies will make sure that AI is being used ethically. To address decision-dependant risks on sustainability, the following decisions should be made; Consequently, transparency, accountability, and explainability in architecture AI systems should be prioritized.

Additionally, the best practices that are associated with the management of AI technology cannot be stagnant due to the advancement in this technology. The scheduling of regular assessments and audits can be strengthened with the possibility of making corrections to AI's impact on sustainable development. Thus, it is reflected that establishing a proper balance between creativity and focusing on the threats and responsibilities of AI makes the achievement of the overall strategic goal possible and contributes to creation of a sustainable future for humanity in terms of conservation of the natural environment and ensuring socio-economic justice.

REFERENCES

Abbas, J., Balsalobre-Lorente, D., Amjid, M. A., Al-Sulaiti, K., Al-Sulaiti, I., & Aldereai, O. (2024). Financial innovation and digitalization promote business growth: The interplay of green technology innovation, product market competition and firm performance. *Innovation and Green Development*, *3*(1), 100111. DOI: 10.1016/j.igd.2023.100111

Alhosani, K., & Alhashmi, S. M. (2024). Opportunities, challenges, and benefits of AI innovation in government services: A review. *Discover Artificial Intelligence*, *4*(1), 18. DOI: 10.1007/s44163-024-00111-w

Cooper, R. G. (2024). The AI transformation of product innovation. *Industrial Marketing Management*, *119*, 62–74. DOI: 10.1016/j.indmarman.2024.03.008

Cooper, R. G. (2024). The AI transformation of product innovation. *Industrial Marketing Management*, *119*, 62–74. DOI: 10.1016/j.indmarman.2024.03.008

Du Plooy, H., Tommasi, F., Furlan, A., Nenna, F., Gamberini, L., Ceschi, A., & Sartori, R. (2024). A human-centered perspective on individual risks for digital innovation management: An integrative conceptual review. *European Journal of Innovation Management*, *28*(11), 49–76. DOI: 10.1108/EJIM-09-2023-0821

Felicetti, A. M., Corvello, V., & Ammirato, S. (2024). Digital innovation in entrepreneurial firms: A systematic literature review. *Review of Managerial Science*, *18*(2), 315–362. DOI: 10.1007/s11846-023-00638-9

Glaeser, S., & Lang, M. (2024). Measuring innovation and navigating its unique information issues: A review of the accounting literature on innovation. *Journal of Accounting and Economics*, *78*(2-3), 101720. DOI: 10.1016/j.jacceco.2024.101720

Hassan, S. S., Meisner, K., Krause, K., Bzhalava, L., & Moog, P. (2024). Is digitalization a source of innovation? Exploring the role of digital diffusion in SME innovation performance. *Small Business Economics*, *62*(4), 1469–1491. DOI: 10.1007/s11187-023-00826-7

Iftikhar, A., Ali, I., Arslan, A., & Tarba, S. (2024). Digital innovation, data analytics, and supply chain resiliency: A bibliometric-based systematic literature review. *Annals of Operations Research*, *333*(2), 825–848. DOI: 10.1007/s10479-022-04765-6 PMID: 35611176

Kamal, R. (2024). Cyborg Employees: Exploring the Benefits and Challenges of Transhumanism in Modern Work Environments. In *Embracing Transhumanism and Genomics in Human Resources Management* (pp. 1-20). IGI Global.

Kamal, R. (2024). Exploring the Synergy: Venture Capital's Impact on the Startup Ecosystem. In *Fostering Innovation in Venture Capital and Startup Ecosystems* (pp. 242-264). IGI Global.

Kamal, R., Kaur, M., Kaur, J., & Malhan, S. (2024). Artificial Intelligence-Powered Political Advertising: Harnessing Data-Driven Insights for Campaign Strategies. In *The Ethical Frontier of AI and Data Analysis* (pp. 100-109). IGI Global.

Kaur, J., Kamal, R., Kaur, M., & Malhan, S. (2024). The Symphony of Efficiency: Transforming Healthcare Through Robotic Process Automation. In *Clinical Practice and Unmet Challenges in AI-Enhanced Healthcare Systems* (pp. 245–261). IGI Global. DOI: 10.4018/979-8-3693-2703-6.ch013

Lodhi, S. K., Gill, A. Y., & Hussain, H. K. (2024). Green Innovations: Artificial Intelligence and Sustainable Materials in Production. *BULLET: Jurnal Multidisiplin Ilmu*, *3*(4), 492–507.

Nahar, S. (2024). Modeling the effects of artificial intelligence (AI)-based innovation on sustainable development goals (SDGs): Applying a system dynamics perspective in a cross-country setting. *Technological Forecasting and Social Change*, *201*, 123203. DOI: 10.1016/j.techfore.2023.123203

Parolin, G., Mcaloone, T. C., & Pigosso, D. C. (2024). How can technology assessment tools support sustainable innovation? A systematic literature review and synthesis. *Technovation*, *129*, 102881. DOI: 10.1016/j.technovation.2023.102881

Salaheldeen, M., & Battour, M. (2024). Fostering innovation capability and sustainable innovation in halal industry: The role of halal entrepreneurs' success. *Journal of Islamic Marketing*, *15*(3), 777–799. DOI: 10.1108/JIMA-12-2022-0323

Sawyer, R. K., & Henriksen, D. (2024). *Explaining creativity: The science of human innovation*. Oxford university press.

Saxena, P. K., Seetharaman, A., & Shawarikar, G. (2024). Factors That Influence Sustainable Innovation in Organizations: A Systematic Literature Review. *Sustainability (Basel)*, *16*(12), 4978. DOI: 10.3390/su16124978

Sengupta, S., Choudhary, S., Obayi, R., & Nayak, R. (2024). Reducing food loss through sustainable business models and agricultural innovation systems. *Supply Chain Management*, *29*(3), 540–572. DOI: 10.1108/SCM-01-2023-0059

Shah, N., Zehri, A. W., Saraih, U. N., Abdelwahed, N. A. A., & Soomro, B. A. (2024). The role of digital technology and digital innovation towards firm performance in a digital economy. *Kybernetes*, *53*(2), 620–644. DOI: 10.1108/K-01-2023-0124

Simeon, I. A., Caballero, R., & Lepez, C. O. (2024, February). Curricular innovation in the knowledge society. In *Seminars in Medical Writing and Education* (Vol. 3, pp. 58-58).

Singh, A., Kanaujia, A., Singh, V. K., & Vinuesa, R. (2024). Artificial intelligence for Sustainable Development Goals: Bibliometric patterns and concept evolution trajectories. *Sustainable Development (Bradford)*, *32*(1), 724–754. DOI: 10.1002/sd.2706

Singh, P. K., & Maheswaran, R. (2024). Analysis of social barriers to sustainable innovation and digitisation in supply chain. *Environment, Development and Sustainability*, *26*(2), 5223–5248. DOI: 10.1007/s10668-023-02931-9 PMID: 36687733

Steinhoff, J. (2024). AI ethics as subordinated innovation network. *AI & Society*, *39*(4), 1995–2007. DOI: 10.1007/s00146-023-01658-5

Vargo, S. L., Fehrer, J. A., Wieland, H., & Nariswari, A. (2024). The nature and fundamental elements of digital service innovation. *Journal of Service Management*, *35*(2), 227–252. DOI: 10.1108/JOSM-02-2023-0052

Wang, H. (2024). Linking AI supply chain strength to sustainable development and innovation: A country-level analysis. *Expert Systems: International Journal of Knowledge Engineering and Neural Networks*, *41*(5), e12973. DOI: 10.1111/exsy.12973

Yanamala, A. K. Y., & Suryadevara, S. (2024). Emerging Frontiers: Data Protection Challenges and Innovations in Artificial Intelligence. *International Journal of Machine Learning Research in Cybersecurity and Artificial Intelligence*, *15*(1), 74–102.

Chapter 2
Intelligent Agriculture

Poornima Pandian
https://orcid.org/0000-0002-5751-0859
Sri Sai Ram Engineering College, India

K. Susithra
https://orcid.org/0009-0004-4302-7315
Sri Sai Ram Engineering College, India

B. G. Panchami
https://orcid.org/0009-0000-4352-8209
Sri Sai Ram Engineering College, India

R. Reshma
https://orcid.org/0009-0005-1353-3863
Sri Sai Ram Engineering College, India

ABSTRACT

Agriculture development is the process in which the conditions are required for fulfillment of agricultural potential. To achieve these conditions, we need the required knowledge and the availability of technology to allocate the inputs and outputs of the system. Agriculture is the major source of livelihood since most of the laborers who are working in the agriculture sector are shifting their work to other sectors by automating agriculture. It is the key driver of international trade, especially for rice. It also reduces

DOI: 10.4018/979-8-3693-8643-9.ch002

rural poverty by increasing the income of the residents. It not only helps to develop the agricultural sector but also supports the manufacturing sector by providing raw materials. Advancement in agriculture with the help of technology provides an increase in economic growth. The moisture level, fertility, and pesticide level of the soil need to be monitored for precision agriculture. Weather prediction and disease identification are also required for efficient agriculture.

1. INTRODUCTION

Since two-thirds of India's population works in agriculture, agriculture is the foundation of the country's economy. Water is essential in the field throughout the growing season in agriculture. These days, the lack of rainfall and other factors like climate change are causing the water level to gradually drop. Therefore, it is important to urge farmers to switch to other fields where groundwater is running low. Long-term food security would be negatively impacted if it continues. Therefore, providing the necessary quantity of water is an essential procedure. An individual keeps an eye on the majority of the irrigation system (P. Pandian et al. 2024).

Plant growth, disease prediction, soil and irrigation management, weather forecasting, and animal management are all areas that use AI approaches and are thought to be important areas of farming. The review suggests future research goals in this area in addition to highlighting the successful application of AI at various layers of smart agriculture architecture. Because of recent technological advancements that can efficiently process large amounts of data and enable timely intelligent decisions similar to human decisions, we found that the deep learning algorithms used in recent studies have performed far better than the conventional machine learning algorithms (S. Qazi et al. 2020).

The AI-powered Agri Agent will offer real-time solutions, connect farmers and buyers on a single platform, and offer price indexing and price prediction (P. Pandian et al. 2024). It is predicted that in the next three decades, there will be 10 billion people on the planet. The demand will be driven by the growing population and rising living standards in cities. By 200 percent for food. Global food security issues will eventually arise from changes in the climate and a reduction in the amount of land available for agriculture. The main social and economic pillars of any country's prosperity and self-sufficiency are agriculture and related sectors. While the majority of farmers still employ traditional farming practices, the majority of organized agricultural linked sectors use sophisticated machinery and technologies. Better use of agricultural land resources and increased output can be ensured by integrating data analysis, technological advancements, and farming forecasts.

The main driver of the current digital revolution is artificial intelligence (AI). Without question, artificial intelligence (AI) has enormous potential to benefit individuals and their surroundings. Artificial Intelligence may surely help find creative solutions to the most pressing issues facing our human civilization in almost every area of life, from the health of every individual to the ecosystems of agriculture and forests that affect our entire globe. One particular aspect is highlighted in this work. Notwithstanding all of its benefits, there is a significant and unforeseen risk of novel, unknown hazards associated with the broad usage of Artificial Intelligence technology. All stakeholders involved must guarantee the deployment of reliable, moral Artificial Intelligence, and improper use of the technology must be prevented (R. Bhati & S. Mittal, 2023).

Robotics and artificial intelligence may be beneficial for societal demands. Agriculture primarily boosts agricultural productivity and output in order to meet the world's expanding population's constant food needs and avoid famine. Artificial intelligence has the ability to suggest effective and workable solutions for the issues and difficulties facing agriculture, such as rising labor costs, increased cultivation costs, crop failures linked to erratic yield because of diseases, erratic rainfall, climatic changes, deterioration of soil fertility, and shifting agricultural commodity market prices (Zha & Jiali, 2020).

In 1983, the use of computers in agriculture was first documented. Various methods, ranging from databases to decision support systems, have been proposed to address the current issues in agriculture. AI-based systems have been discovered to be the best performing among these options in terms of accuracy and resilience. Situations in the dynamic field of agriculture cannot be generalized to imply a shared answer. AI methods have made it possible for us to comprehend the finer points of any scenario and offer the best answer for that specific issue. As different AI techniques are developed, increasingly complex issues are being resolved (Bannerjee et al. 2018).

Humans have pushed the boundaries of our cognitive abilities with the introduction of technology in this digital age, and we are attempting to combine the natural brain with an artificial one. The field of artificial intelligence was born out of this ongoing investigation. It is the method by which a human can create a machine with intelligence. AI is a branch of computer science that can identify its environment and should flourish in order to achieve the highest possible success rate. AI needs to be able to perform tasks based on prior knowledge. Some fields that improve machine performance and aid in the development of more sophisticated technologies include deep learning, CNN, ANN, and machine learning (Jha et al. 2019).

Agricultural tasks that once required days of human sweat and draft animal labor were reduced to a few hours of work thanks to mechanization. In emerging nations like India, this can be seen as the initial degree of automation that changed agricultural tasks. India's agricultural mechanization is still in its infancy and is expanding at a rate of 7.5% annually; as digital technologies progress, this will become more intelligent and quicker. Large-scale labor migration is one of the primary problems of the modern era. Researching the workforce in Indian agriculture revealed that the proportion of workers in this sector fell from 59.1% in 1991 to 54.6% in 2011, and it was predicted to reach 40.6% in 2020 (Subeesh & Mehta, 2021)

The goal of artificial intelligence (AI), a subfield of computer science, is to create intelligent tools and algorithms that enable computers to simulate human intelligence. The development of intelligent software, devices, and systems based on an understanding of how the human brain thinks, learns, makes decisions, and works to solve problems is what led to this creation. Similar to the human brain, these intelligent gadgets receive training data and, for each valid input, provide the appropriate outcome. Today's agriculture is equipped with robots and drones that are programmed to collect data from production fields in order to develop precision agriculture, often known as "smart farms" (Vijayakumar et al. 2022).

Many technological approaches have been employed and suggested in recent years due to the growing importance of smart systems in sustainable agriculture. Through the use of cutting-edge agricultural technologies, smart farming maximizes intricate agricultural systems. With less money, irrigation, and manpower, it seeks to grow and harvest higher-quality crops. Generally speaking, harvesting and plant growth are important aspects of smart farming. Intelligent mechanism models will therefore be highly helpful in resolving issues and accomplishing objectives. However, it's crucial to take a comprehensive approach that addresses the entire process rather than just one or two particular goals. An application model that encompasses the entire process, from growing to harvesting, is put out in this study. Any farmland can be used to cultivate a variety of crops using this strategy. This is accomplished by combining hardware and software techniques, enabling their practical application (Kiani et al. 2022).

The conventional techniques that farmers employed were insufficient to meet these demands. New automated techniques were presented. In addition to meeting food needs, these innovative techniques gave billions of people job opportunities. Agriculture has undergone a revolution thanks to artificial intelligence. Crop yield has been shielded by this technology from a number of concerns, including population expansion, climate change, joblessness, and issues with food security. With the aid of sensors and other tools built into robots and drones, it will examine the numerous uses of artificial intelligence in agriculture, including irrigation, weeding, and spraying. These technologies reduce the need for excessive amounts of water, pesticides, and herbicides, preserve soil fertility, aid in the effective use of labor, increase productivity, and enhance quality (Tanha Talaviya et al. 2020).

Farmers have recently shown a great deal of interest in smart agriculture approaches. The growing availability of low-cost, Internet of Things (IoT)-based wireless sensors that can remotely monitor and report crop, field, and climate conditions is one of the main causes of this. This makes it possible to manage resources effectively, such as reducing the amount of water needed for irrigation and the usage of harmful pesticides. In order to reduce crop illnesses and insect infestation, farmers may also be able to use autonomous farming equipment and improve their forecasts of the future by analyzing historical and current data. These two enabling technologies have transformed traditional agricultural methods when combined (S. Qazi et al. 2020).

2. ROLE OF AI IN AGRICULTURE

Precision agriculture encourages data-driven decision-making, which supports our choices and provides additional information to help us test and choose the most effective agricultural stage. The development of precision agriculture has been greatly aided by remote sensing, multispectral data from satellites, particularly because of advancements in sensors that provide spatiotemporal information that enables farmers and researchers to survey, monitor vast areas of agricultural land, and take well-informed decisions to maximize field health and productivity (V. Singh & G. Singh, 2023).

Big data, Artificial Intelligence and Machine Learning technologies can be used by businesses to predict prices, determine tomato crops and output, and spot pest and disease infestations. They can give farmers advice on future pricing trends, crop varieties to plant for maximum profit, pesticide use, and demand levels. Artificial Intelligence significantly lessens the labor and resource deficit, can serve as a powerful tool that organizations may use to address the increasing complexity of modern agriculture (Mohd Javaid et al. 2023).

One of the issues that has been identified is that modern technology can help people in their daily life. One is an AI-enabled bot. Any service sector can use these AI-integrated bots. It presents AI-enabled real-time question and response platforms for the agricultural sector, giving farmers the ability to find out information about the queries they occasionally pose to agrarian officers. Many questions about farmer's fields and crops are common when thinking about the agriculture industry. To find solutions to their issues, farmers ought to have waited for agricultural officers (L. A. Imalka et al. 2022).

By employing wireless sensor networks, agricultural robots (AGBOTS), and drones to measure, monitor, and detect agricultural parameters (such as temperature, humidity, moisture, yield, and crop fertility), smart agriculture ensures agricultural automation. Smart agriculture uses wireless sensors to ensure agricultural automation. Systems for measuring, tracking, and identifying agricultural parameters (such temperature, humidity, moisture, yield, and crop fertility) include networks, agricultural robots (AGBOTS) and drones (F. K. Shaikh et al. 2020).

Weather, crop knowledge, soil parameters and other factors all have a significant impact on the quality and production of products. If these specifications are created whether through automated measurements, openly accessible data, or past knowledge, farmers can be better equipped to make decisions. Crop output and quality can be raised at the farm level through data-driven inputs and decision-making. In order to automate the farming process, smart agriculture combines sensors, publicly accessible data, and existing expertise with machine intelligence and connection.

Farmers need to adopt technology solutions to close the gap between the supply and demand of agricultural products. The majority of people in the traditional farming community are not knowledgeable enough to use the new technologies. In order to close the gap between supply and demand, the country must provide a framework that allows farmers with better technological solutions to increase productivity.

Farmers can be empowered for improved outcomes if these criteria are made available by automated measures, publicly accessible data, or prior knowledge ability to make decisions. Crop output and quality can be raised at the farm level through data-driven inputs and decision-making. In order to automate the farming process, smart agriculture combines sensors, publicly accessible data, and existing expertise with machine intelligence and connection (P. A & P. Kumar Reddy, 2022).

3. MANAGEMENT IN AGRICULTURAL PRACTICES USING ARTIFICIAL INTELLIGENCE

3.1 Soil Management

A number of factors are utilized to look at the land's suitability for farming. These parameters aid in estimating the amount of water and nutrients needed. They are necessary for the development of superior crops. Nitrogen (N), organic carbon (OC), and other nutrients are typically determined through laboratory testing. They are time-consuming and expensive. Farmers are now able to make logical decisions because of the application of Artificial Intelligence and Machine Learning approaches (F. K. Shaikh et al. 2022)

In contemporary agriculture, crop and soil monitoring—which entails continuously assessing crop and soil conditions—is essential. Farmers may make decisions that are more informed, examine how resources are being used, and increase agricultural productivity while reducing their negative effects on the environment (R. Rani et al. 2023).

3.2 Irrigation Management

One of the key elements that determines the crop's health and provides nutrients is proper irrigation. The problem with irrigation is poor water supply management, which results in crops using less water. Other issues in agriculture include the usage of antiquated irrigation techniques and a dearth of contemporary techniques like surface irrigation, drip irrigation, sprinkler irrigation, etc. (V. Singh & G. Singh, 2023)

IoT-based AI in irrigation has recently offered farmers some degree of control over irrigation management. There is also growing interest in using remote sensing techniques to forecast crop and field water needs. However, because spatial and temporal maps lack sufficient resolution, the use of remote sensing technology in irrigation management has been limited.

3.3 Pest Management

In order to keep crops healthy and guarantee agricultural productivity, pest control and disease diagnostics are essential. It includes methods and approaches to manage and lessen the effects of pests, such as weeds, insects, and other creatures that harm crops. Crops must be regularly inspected for indications of damage and insect presence. It's critical to identify the precise bugs generating issues and comprehend their life cycle. Crop rotation, companion planting, timing plantings to interfere with insect life cycles, and maintaining appropriate plant spacing and hygiene to minimize pest habitats are all beneficial practices.

3.4 Precision Rate Spraying

The majority of agricultural output around the world still makes use of robotics' innovations and advancements. Almost all agricultural industrial activities, including planting and seeding, caring for crops, harvesting, and processing, use these robots. The creation and application of these robots can be seen in a variety of settings, such as open fields, buildings, greenhouses, and orchards, which are home to a wide variety of plants (K. Lochan et al. 2024).

3.5 Weather Forecasting

Numerous websites and apps for weather give current weather predictions for particular areas. These tools are useful for farmers to monitor daily and weekly meteorological parameters, such as wind speed, humidity, precipitation, and temperature. Real-time weather information particular to the farm can be obtained by installing weather stations there. Farmers can precisely monitor the conditions of their farms thanks to these stations, which detect temperature, humidity, wind direction, and precipitation. A more comprehensive picture of weather patterns is provided by satellite and radar data, which can also be used to forecast severe weather occurrences like storms or periods of intense precipitation. Farmers that use commercial radar imagery providers or government agencies can obtain this data.

3.6 Yield Efficiency

Since the majority of farming products are produced by natural forces, farms and farmers are under a great deal of stress. The erratic nature of the rain, a shortage of labor, and the yearly requirement for higher yields all contribute to this stress. This implies that in order for us to almost reach our goals, the agricultural sector will need to grow significantly during the next several years, and farm productivity will need to double (Mohd Javaid et al. 2022).

3.7 Predicting Crop Yield

One of the hardest things in agriculture is forecasting crop yields. It is crucial to decision-making on a field, regional, and worldwide scale. Crop yield is predicted using crop, soil, weather, and environmental factors. Numerous models for decision support are utilized to extract important crop characteristics in order to make predictions. In precision farming, observation (sensing technologies), variable rate technologies, management information systems, and cropping system responses to intra- and inter-variability are all crucial. Precision agriculture's advantages entail lowering the impact on the environment while raising crop yield and quality. Crop yield simulations aid in understanding the combined effects of pests, illnesses, nutrient and water shortages, and other field variables during the growing season (Nyéki, A & Neményi, M, 2022).

3.8 Plant Diseases

Cutting-edge technologies have played a major role in the notable advancements in plant disease diagnosis, monitoring, and management during the past 15 years. Advanced technologies including optical sensors, artificial intelligence, microsensor networks, and self-driving cars have propelled recent developments in precision agriculture. In contrast to the conventional, uniform treatment of vast crop areas, these technologies have made it possible to create innovative cropping systems that allow for tailored crop management. In this field, research is typically an interdisciplinary, highly collaborative endeavor. To create all-encompassing solutions, it brings together professionals from a variety of disciplines, including plant pathology, computer science, statistics, engineering, and agronomy. Despite the gains, integrating automation or improvements in decision-making accuracy into agricultural practices is still a challenge (Mahlein & Anne-Katrin, 2024).

4. TECHNOLOGY USED

To help and automate the agricultural models, a clever artificial intelligence-based Internet-based solution is suggested. The platform, which is built on artificial intelligence, links many Internet of Things advancements in the agriculture sector. A business-to-business strategy will be used to implement the combined IoT-AI stage. The idea permits the creation of interconnected systems and equipment from several manufacturers to exchange data. A group of agricultural equipment manufacturers must work together based on a data idea to support this comprehensive system and establish autonomous agricultural systems as the new norm. Data from computers and other devices is fed into a cloud application, which is subsequently shared for use by other Internet of Things users. Every Internet of Things component should have a unique set of data recovery and redirection rules because not all data is intended to be shared (A. R. Reddy et al. 2023).

4.1 Soil Management:

Using AI-powered satellites and unmanned aerial vehicles (UAVs) to map soil variability is a cutting-edge approach to soil management in contemporary agriculture. Farmers can administer pesticides, fertilizers, and water more precisely thanks to AI-powered precision agriculture tools, which also lessen waste and their negative effects on the environment. Artificial Intelligence can maximize input use for sustainable soil management by evaluating crop performance metrics and soil data (Eli-Chukwu & Ngozi Clara, 2019).

Based on past data, AI models—such as machine learning algorithms—can forecast the characteristics of soil and health indicators. In order to increase soil health and productivity, this can assist farmers in making well-informed decisions regarding crop selection, fertilizer use, and irrigation techniques. Some of the AI models are used in soil management systems. A set of created plausible management options, a simulator that assesses each alternative, and an evaluator that decides which alternative satisfies the user-weighted multiple criteria make up management-oriented modeling (MOM), which reduces nitrate leaching. Through the use of "best-first," a tactical search method, and "hillclimbing," a strategic search method, management-oriented modeling (MOM) determines the shortest route between start nodes and goals. The Soil Risk Characterization Decision Support System (SRCDSS) requires engineering competence in three stages: knowledge collection, conceptual design, and system implementation. An artificial neural network (ANN) model predicts the soil texture (sand, clay, and silt concentrations) based on hydrographic data from a digital elevation model (DEM) and features from existing coarse-resolution soil maps (Shalini, 2024).

4.2 Irrigation Management:

The authors (Hinnell et al.2010) have introduced the Neuro-Drip, an irrigation system based on neural networks. Together with the Internet of Things (IoT), Artificial Intelligence (AI) has made significant contributions to this subject. Information about soil moisture content, plant water content, atmospheric humidity, temperature, and other factors must be gathered in order to implement smart irrigation. Soil moisture sensors, temperature sensors, humidity monitoring sensors, and other devices can be used to collect this data. These sensors are linked to inexpensive Arduino-based systems that store the collected data and run analysis algorithms to forecast the crop's water needs at a specific moment (Sinwar et al. 2020). These sensors assess the moisture content of soil at different depths using technology based on capacitance or resistance. To record moisture gradients and offer information on the water availability in the root zone, several sensors are positioned at varying soil depths (Benhmad et al. 2024).

The suggested framework's communication protocols are adaptable and transparent enough to support both wired and wireless approaches. The communication between different components in the edge-IoT system necessitates intensive probing using the variety among the devices in network edges in order to maximize the potentiality in edge computing components. The irrigation systems require a communication technology like Zigbee to convey the data collected from pivot sensors (Premkumar & Sigappi, 2022).

5. PEST MANAGEMENT:

5.1 Data Augmentation:

Image augmentation to increase the pests training sample database is one of the methods for pest detection. The two main categories of data augmentation techniques are geometric transformation and photometric transformation techniques. Three changes to the geometry (rotating, flipping, and cropping) and three changes to the luminance (fancy principal component analysis, edge enhancement, and color jittering). The Convolutional Neural Network (CNN) model is trained using both the original and amplified image samples as training data. The testing findings demonstrated that every amplification technique raises CNN's accuracy, with cropping having the most noticeable effect. Thus, the authors hypothesize that the training samples produced by the cropping technique will enhance CNN performance by preventing over-fitting of the data (Chen et al. 2020)

5.2 Image Pre Processing:

This article's automatic plant pest identification and recognition technique makes use of the correspondence filter and k-mean clustering. We use a variety of plant pests that we have gathered from various internet sources, including Alabama Nursery and Landscape, as a test bed. There are two stages to the suggested pest detection and recognition algorithms: the detection stage and the recognition stage. In order to develop the color transformation structure for the RGB plant pest photos, the algorithm first gathers digital photographs from the surroundings and loads them as filter construction images (Miranda et al. 2014).

5.3 Sensor -Based Canopy Measurement Technologies for Precision Spraying:

As mentioned in the role of agriculture, precision spraying is used in tree canopy parameter measurements that call for many sophisticated sensor technologies and applications. The subsequent sections provide a general overview of the use of various sensors for accurate spraying.

Range-sensing systems employ ultrasonic sensors, while machine vision systems use camera sensors (RGB, NIR, hyperspectral, multispectral, and infrared) and laser sensors to gather information about tree canopy and thoroughly assess canopy targets and attributes (Mahmud et al. 2021).

6. WEATHER FORECASTING:

An ANN is modeled after a biological neural network. This model is viable from an economic standpoint. The artificial intelligence-based model has the advantage of providing a better solution for many problems without requiring intricate mathematical calculations between the variables. Learning from the data is the overarching motivation for using the tool. Three layers make up an ANN: the input layer, the hidden layer, and the output layer. Complex nonlinear statistical data between inputs and outputs is present in the network. An artificial neuron is capable of gathering signals, processing them, and sending signals to other neurons. Applications of artificial neural networks (ANNs) include signal classification, image processing, face identification, prediction and forecasting, system identification and control, and more (Meenal et al. 2022).

The Machine Learning (ML) model was created to ascertain the soil's present moisture and nutrient content level as well as forecast its future levels. The model adhered to the standard procedures used by all machine learning algorithms. In order to prepare for training, the data was first gathered and cleaned. It was then transformed into JSON format and uploaded to the cloud-based machine learning platform. Additionally, the data was divided into training and validation sets. After that, the data underwent preprocessing to produce training features. Following pre-processing, a machine learning model was created, and the soil's moisture content and nutrient content served as the model's training inputs. For the purpose of classifying whether the soil is under stress or has enough nutrients for sound decision-making, the data was trained and validated to create a Machine Learning (ML) model (Pierre et al. 2023).

7. YIELD EFFICIENCY:

7.1 Random Forest:

Learning algorithms that build a group of classifiers rather than a single classifier and then vote on their predictions to classify incoming data points are known as ensemble classification methods. RF, bagging, and boosting are recent ensemble classifiers. In the algorithm for bagging, to develop a classifier, numerous bootstrap samples are selected from a training data set with replacement. A tree is built for each bootstrapped sample so that subsequent trees are built independently from previous trees, and a simple majority vote is used for prediction. By using iterative re-training, boosting, on the other hand, increases the weights of samples that are mistakenly classified throughout the course of rounds, making them more significant in subsequent iterations. Although technique has certain drawbacks, boosting typically lowers the classification's variance and bias and is typically far more accurate than bagging. It is sensitive to noise, sluggish, and capable of overtraining (Nyéki & Neményi, 2022).

7.2 Decision Tree Algorithm:

For the separation of discrete class data for land cover categorization, decision trees—supervised organization schemes based on recursive binary divisions that adhere to numerous updated guidelines—have emerged as a compelling substitute. After receiving a large number of inputs, a decision tree arranges tests to produce an outcome. In order to construct the instruction, trees divide areas that are more homogeneous with respect to their class variable using recursive binary division. Decision Tree classifiers provide multivariate designs that rely on numerous decision instructions that are implemented at each test node and are defined by a variety of linear discriminant equations and parameter combinations (Ok et al. 2012).

7.3 SVM Based Calculation:

SVM converts a set of non-linear decision boundaries in the original dataset into linear boundaries of a higher-dimensional construct using a user-defined kernel function (Han et al., 2007). Support vector machines (SVMs) are supervised learning algorithms that are regarded as heuristic algorithms that are based on statistical learning theory. Finding the best hyperplane to divide two classes is the goal of SVMs for classification. Training data is used to identify the ideal hyperplane, and validation data is used to confirm that it can generalize. Therefore, the main goal of SVM is to identify the optimal separating hyperplane (OSH) among all potential hyperplanes. This is done by solving an optimization problem using quadratic programming techniques and Lagrange multipliers (Pal and Mather, 2004). The five internal parameters—

a) kernel function, which included linear, polynomial, radial basis function, and sigmoid kernels;
   (b) bias in kernel function, which applied only to the polynomial and sigmoid kernels;
   (c) degree of polynomial function, which applied only to this type of kernel;
   (d) Gamma in kernel function, which excluded the linear kernel; and
   (e) error penalty—were carefully adjusted to create a total of 53 distinct support vector machines.

An example for SVM model is in the ARIMA–SVM model, the SVM predicts the ARIMA model residual, incorporating the nonlinear rule into the SVM's prediction outcomes. Analyzing the model's prediction accuracy can be aided by residual estimation.

8. PREDICTING CROP YIELD:

Using yearly Sentinel-2 remote sensing pictures from China, Wang et al. present the dual-path interactive network (DPIN), a deep learning-based method for mapping paddy rice on a wide scale. The method employs opaque and cirrus cloud pixels for cloud detection and unsupervised learning models for crop mapping. A neural network model, a simple training model, and a related training model were used to estimate the biomass. RTM outperformed STM (RMSE = 1.226 kg/m2), as indicated by the Root Mean Square Value (RMSE) of 0.816 kg/m2. According to the investigation, AI systems achieved over 90% accuracy in mapping rice farms and estimating yields. This is a major agricultural milestone in terms of rice yield projections. Nevertheless, the suitability of these models depends on the type of rice cultivated in a particular area. The temporal heterogeneity of the rice production cycle in a given location also contributes to misclassification in rice crop mapping and prediction.

9. PLANT DISEASES:

9.1 Machine Learning in Classifying Plant Diseases:

The development of blister blight disease is correlated with environmental factors. Rainfall has a negative correlation with blister blight disease growth, although temperature and humidity have a strong positive correlation. The regression line model is employed for implementation since there is a correlation between the predictor and response variable.

9.2 Image Processing in Plant Diseases:

Two algorithms were used—one for training various SVM classifiers and another for testing these various SVM classifiers—are suggested. Image segmentation is done using K-means clustering. Both methods involve image acquisition, where eight pictures of the chili plants were taken using an 8MP Raspberry Pi Camera V2. The photos can be taken at varying heights and with the plant facing the camera in varied orientations thanks to a pre-programmed movement sequence between the camera and the plant. After that, MATLAB stores these pictures for later examination. It has 233 pictures of leaves with intricate backgrounds. Additionally, the plant level is a component of the Plant Identification Database. depicts the visual comparison between the current model and the Adaptive Snake approach; the original image is shown in the first column, the existing model's Region of Interest is shown in the second, and the third column shows the ROI detection using the adaptive snake technique, which is the suggested model. When we look at each image in this case, we can see that ASA accomplishes absolute segmentation and superior segmentation over the current one.

9.3 Internet of Things in Plant Disease Deduction:

Weather, precipitation, sediment, and pH sensors are reliable sources of information. Periodic data sources are used in the collection of historic data. Big data and responses from local producers make up this data set. The qualifications for specialized fields are divided into an effective algorithm technique for categorizing results, together with the application of identification technologies. For farmers, incorporating machine learning or statistical models into an economy centered on agriculture will be extremely advantageous. This strategy may help with both production gains and suitable farming practices. When it comes to identifying plant diseases, precision farming and plant phenotyping both have unique requirements and difficulties. New methods must be developed and integrated into established monitoring and grading systems in order to achieve objective and trustworthy automated diagnosis and detection of plant diseases. Optical sensors show promise as noninvasive diagnostic and disease detection techniques. A growing range of noninvasive sensors and imaging technologies are available to aid in plant disease detection and diagnosis. Precision agriculture and plant phenotyping have new prospects thanks to advancements in sensor and information technology and the growth of geographic information systems.

9.4 Fusion Approaches in Plant Disease Diagnosis:

Amin et al. categorize good and diseased corn plants using a feature fusion end-to-end deep learning technique. It makes use of DenseNet121 and EfficientNetB0, two pre-trained CNN techniques. The model then learns more about the dataset by combining the characteristics with additional methods. The outcome is then contrasted with ResNet152 and

InceptionV3, two more trained CNN models. While ResNet and InceptionV3 have accuracy rates of 98.37% and 96.26%, respectively, they attain an accuracy rate of 98.56%. Accuracy increases when features taken from various CNNs are combined to create a more complicated set.

10. CHATBOTS:

10.1 Natural Language Processing:

A theory-driven collection of computer methods for the automatic analysis and representation of human language is known as natural language processing, or NLP assert that speech research should not be confused with NLP research recognition but focuses on the symbolic meaning and interface adjustments required after word recognition. Actually, to develop real verbal conversations with computers, a language processing program must be combined with a speech recognition system. According to the researchers, NLP research is undergoing a paradigm change and is no longer predicated on methods for recognizing and comprehending loose words. However, we now start to more regularly research semantic strategies, which the authors refer to as a jump from the syntactic curve to the semantics curve. Eventually, we will reach the pragmatics curve, where computer programs will be able to explore and construct whole narratives.

10.2 Pattern Matching:

Chatbots organize messages and generate pertinent user responses using a pattern matching technique. A common structured model called artificial intelligence markup language (AIML) is used to develop user interfaces that are easy for humans to understand and respond to in the right format.

10.3 Natural Language Understanding:

An online chatbot is a clever piece of software designed to mimic human communication when dealing with users online. To determine user intent, natural language processing and keyword extraction are used. One crucial piece of technology that determines whether chatbots succeed or fail is natural language understanding. For entity extraction, response retrieval, and intent categorization in chatbots, this work employs Rasa NLU in conjunction with machine learning. Social media platforms, which offer simple-to-integrate webhooks to support the creation of intelligent apps, are popular places for chatbots.

11. WEED MANAGEMENT:

To lessen the yield drop, farmers use a variety of weed control techniques. However, the majority of current methods (16 Phyton, 2024) rely on chemical herbicides, which poses significant environmental dangers and speeds up the growth of herbicide-resistant weeds. The creation of automated weed control tools for real-time field management has drawn interest because of the shortage of labor, the high cost of human weeding, and the growing demand for organic food. For effective weed control in large-scale agriculture, labor- and money-saving methods are required. One great strategy for guaranteeing the sustainability of agricultural production systems is automated weeding. Although herbicides are frequently used in agriculture to manage weeds, their use pollutes the environment and puts those who apply them in danger of poisoning. The pre-processed output undergoes segmentation, which groups related pixels to create a connected object with uniform characteristics. This entails gathering only the vegetation portion and removing the plants from the soil and other backdrops. The following step, known as feature extraction, removes the segments' key characteristics. These consist of spatial settings, visual textures, spectral characteristics, and biological morphology. An increase in dimensionality is the outcome of extracting and combining different features. Only the most important feature combinations are chosen for efficacy utilizing a variety of methods, including Principal Component Analysis (PCA), Linear Discriminant Analysis (LDA), Stepwise Linear Discriminant Analysis (SWLDA), etc. Either traditional machine learning-based classification or deep learning-based algorithms can be used for the final classification of weeds and crops.

11.1 Machine Learning:

In agriculture, machine learning algorithms are being utilized more and more to detect, evaluate, and forecast many facets of crop production, such as weed control. Numerous applications for these algorithms exist, including decision support systems, herbicide efficacy prediction, weed species identification, and weed density estimate. Farmers and other agricultural professionals may minimize their influence on the environment, maximize the use of herbicides, and increase the accuracy and efficiency of weed control by utilizing machine learning. In order to identify patterns and connections between input variables (such weed features and environmental factors) and output variables (like herbicide efficacy or weed presence), machine learning algorithms are trained using these data sets. After training, the models can be used to give real-time forecasts or suggestions for weed control tactics. These models' forecasts or suggestions can be used to inform farmers about the best weed control tactics for their specific circumstances. Therefore, using AI-driven models can assist farmers in managing weeds more effectively.

11.2 Support Vector Machines (SVM):

Weed identification is a common use of SVMs. Using hyperspectral data, Karimi et al. used SVM to identify nitrogen stress and dominating grassy weed in corn. There are 72 thin bands in the data, with a range of 408.73 to 947.07 nm. Weed control was the primary factor taken into account, with three nitrogen application rates serving as sub-factors. When nitrogen rates and weed infestation levels were combined, the classification accuracy of the used SVM model with RBF kernel was 69.2%; however, when weed control and nitrogen rates are taken into account independently, the classification accuracy increases to over 80%.

Using LBP to extract textural information, Ahmed et al. presented a texture-based weed classification method. Weeds in the grassland and broadleaf groups were examined. Ten-fold cross-validation was applied to the 200 color photos (100 of each category) in the dataset. The dataset was then randomly divided into 10 subsets, with nine of them being used for training and one for testing. An encoded representation of the image was created by using the LBP operator to calculate the LBP code for every pixel in the input image. This was utilized to create a histogram, which served as the feature vector and represented the textural information of the image. Classification using SVM with RBF kernel yielded a 98.5% classification accuracy.

11.3 Remote Sensing Weed Detection:

Evelyn Pruitt originally used the term “remote sensing” in the late 1950s, and since then, other scientists and academics have attempted to explain its exact meaning (Fussell et al., 1986). In order to translate the images into a format that is legible by humans, remote sensing involves indirect data collecting followed by data manipulation operations like pixel transformation or pixel clustering. RGB-based orthomosaic weed detection entails flying missions and producing geo-registered imagery that, when projected onto the earth's surface, shows the actual position (bird's-eye view of a field). Two distinct approaches are currently used to process an orthomosaic for weed detection. These approaches rely on ML and DL-based techniques. Creating a classifier trained on a single orthomosaic segment is a popular strategy when using machine learning (ML) techniques to distinguish weeds from crop plants. The remainder of the entire orthomosaic is used to further confirm this classifier.

11.4 Deep Learning:

One significant area of machine learning is deep learning (DL). DL algorithms offer numerous benefits over conventional ML techniques for image classification, object detection, and recognition. Crops and weeds can be similar, making it challenging to extract and choose distinguishing traits using machine learning techniques. Because of their powerful feature learning capabilities, DL methods can effectively handle this issue. Few review articles have been written on DL-based marijuana recognition, despite the fact that a large number of research articles have recently been published on the subject. More recently released a review paper that focused on autonomously classifying weeds in crops using point spectroscopy, RGB, and hyperspectral imaging. With few references to more current work, the majority of the articles in this review have used conventional machine learning techniques.

12. RISK MANAGEMENT:

A common tool for examining research trends is bibliometrics, a quantitative research method (Lund, 2022). The metadata of publications taken from the Web of Science (WoS) database, which supplied the prerequisites for developing the study of the topic of risk management in the age of artificial intelligence in agriculture, was used to create this particular bibliometric analysis. As a result, in May 2023, the following criteria were used to query the WoS database: Without using an age filter, TOPIC: ("artificial intelligence") AND TOPIC: (risk) AND TOPIC: (agricultural). After the query was conducted, only articles with these word structures in their titles, abstracts, or keywords were shown. The conceptual basis for obtaining bibliometric research findings based on scientific publications unique to the subject of risk management in the age of artificial intelligence in agriculture is represented by combining these three components in a single query. 348 publications with the word structures "risk management," "food," and "chain" in their titles, abstracts, or keywords were found using the WoS query.

13. APPLICATIONS

Agriculture has changed from being a low-tech industry to high-tech customs for the previous 20 years. More processes are being automated to reduce the need for human and material resources while enhancing economic viability. The United Nations Food and Agriculture Organization projects that by 2050, there will be a 20% increase in population and a 4% decrease in agricultural area due to urbanization. Along with the intensity of production, agricultural production's interaction mechanisms have grown more complicated. Farmers marketed a variety of products directly to consumers and wholesalers, cared for livestock and crops by hand, and even prepared crops and supplies for the next year up until the

1970s. This closed-loop design can no longer be used. The majority of farmers specialize in raising specific products, such uncommon grains or animals, and they do so in artificial settings like sizable animal farms and agricultural systems. Figure 1 shows some uses of AI in the field of agriculture.

Figure 1. Applications of Artificial Intelligence (AI) in agriculture.

13.1 Yield Prediction:

By utilizing a variety of data sources, machine learning strategies, and sophisticated analytics, artificial intelligence (AI) can be applied to agricultural production prediction. Crop output is predicted by Artificial Intelligence (AI) and is dependent on a number of variables, including weather, soil, micro and macronutrient levels, and water content. These components are also required to monitor crop health (Ulfa, et al. 2020).

This paper (Ulfa, et al. 2020) uses 48 wheat yield monitor maps from 23 fields to develop linear-mixed models for forecasting yield. The models were tested using leave-one-field-out cross-validation. The authors investigated multiple vegetation indexes and found minimal differences in their performance. This research clearly demonstrates that predictions based on longer-term average yields are generally more accurate than yield predictions for individual years.

13.2 Yield Mapping:

Yield mapping is a tonal or shaded guide that depicts the scope of yield within an area. Grain fow sensor (determines grain volume gathered), grain moisture content sensor (remunerates for grain moisture variability), GPS antenna (receives satellite signal), Yield screen show with a GPS receiver (geo-reference and records information), header position sensor (distinguishes estimations logged during turns), travel speed sensor.

13.3 Harvesting:

This paper (Kiani et al. 2022) said that the goal of the harvesting phase is to develop a roadmap for each autonomous robot. At this point, the SCSO and related algorithms may not be very reasonable, as our goal is to generate paths and discover a decent route. As a result, a novel solution was developed, utilizing one of the greatest metaheuristic algorithms (ACO) for this topic. Based on this plan, each robot will begin its work of gathering (harvesting) crops in the most effective way. In other words, using this path, each robot collects crops efficiently, resulting in the highest profit. Furthermore, this goal is met in the shortest amount of time feasible while using the least amount of power.

13.4 Automated Pest Control:

Crop destruction caused by pest infestations can result in large financial losses. The environment and beneficial insects may be harmed by the widespread use of pesticides in traditional pest control techniques. AI-powered solutions provide a more sustainable and focused method of pest management. Machine vision and pattern recognition are used by AI-powered pest management systems to identify pests early and enable prompt responses. AI-enabled autonomous drones and robots are able to recognize and target certain pests, precisely spraying pesticides where they are needed. This focused strategy preserves beneficial insects while using fewer chemicals. Additionally, predictive models predict insect outbreaks, which aids farmers in risk management.

13.5 Managing Risk:

To monitor plant growth and detect plant diseases, a variety of plant sensors have been created. UAVs can currently be used in agriculture for crop health monitoring (nutritional diseases, insect/disease damage), irrigation equipment monitoring, weed identification, cattle and animal monitoring, and disaster management. This aids in spotting the issue before the plant exhibits outward signs of an illness or deficit

13.6 Self-Driving Tractors:

AI, self-driving tractors, and the Internet of Things could help address one of the most pressing problems facing farming: a labor crisis. These technologies may also be less costly since they are more accurate and generate fewer errors. Precision farming is made possible by the convergence of AI, IoT, and self-driving tractors. Agricultural robots already do labor-intensive tasks like gathering fruits and vegetables and trimming lettuce. They may operate for longer periods of time, are more accurate, and are less likely to make mistakes. AI developments could revolutionize the sector by offering everything from predictive analytics to deep data insight.

13.7 Plant Health Monitoring:

AI in agriculture can also keep an eye on the sustainability and health of plants. AI, for instance, is capable of identifying illnesses, pests, and soil nutrition on farms. An AI system can determine the areas that require fertilizer. This can significantly lower the amount of fertilizer needed. Food sorting has been transformed by AI and ML, which minimize manual work and guarantee that premium products are sorted precisely. AI-based systems can reduce waste and meet quality standards by sorting according to size, weight, and color. Automation, which reduces manual work, can save businesses money. Examples of this technology include x-ray scanners, lasers, cameras, and robots.

13.8 Detecting Smallest Insects:

They have to make sure that food is produced responsibly and deal with the implications of climate change on a daily basis. To make the greatest choices in terms of the economy, the environment, and the law, several requirements and factors need to be taken into account. This is the exact situation in which digital agriculture may assist. AI systems are able to identify tiny insects. Farmers are able to take the required safeguards since notifications are instantly sent to their smartphones when an intruder is detected. Critical crops may be produced, harvested, and sold more effectively with AI in agriculture.

13.9 Chatbots for Farmers:

Chatbots are conversational virtual assistants that automate conversations with users. Chatbots powered by AI and machine learning can interpret natural language and provide a more tailored user experience. The facility caters to retail, travel, media, and agriculture. Farmers have benefited from receiving answers to their inquiries, guidance, and recommendations.

13.10 Skills and Workforce Evolution:

The authors (Panpatte 2018) proposed that using artificial intelligence, we can build smart farming practices that reduce farmer losses while increasing productivity. Using artificial intelligence platforms, large amounts of data can be gathered from government and public websites, or real-time monitoring of various data can be done using IoT (Internet of Things) and then analyzed with accuracy to help farmers address all of the uncertain issues they face in the agriculture sector. By 2050, the UN predicts that two-thirds of the world's population would reside in cities, diminishing the rural workforce. New technologies will be required to alleviate farmers' workloads: operations will be performed remotely, processes will be automated, risks will be detected, and concerns will be resolved.

13.11 Supply Chain Management:

Regardless of who works in agriculture, it is not always possible to predict precisely how much and when a certain crop will be available. This could lead to significant supply issues when paired with changing consumer demand. Some of the issues that arise in the supply chain may be resolved by data-driven AI solutions. because it makes it easier to see the crops and harvest every season. This is true not only for farmers but also for distributors, packagers, retailers, and other supply chain participants. The data generated by AI will help progress these businesses.

13.12 Early Warnings:

AI applications in plant disease detection present game-changing chances to enhance agricultural productivity, sustainability, and resilience in the face of emerging disease threats. Early warning systems allow for prompt interventions, such as targeted fungicide spraying or quarantine measures, to contain disease outbreaks and minimize economic losses in impacted regions. Farmers, researchers, and policymakers may improve disease surveillance, diagnosis, and management techniques by utilizing AI-driven technology, which will ultimately improve global food security and livelihoods.

13.13 Monitoring and Storage Control:

Aside from diseases and pests, other crucial elements of agriculture include crop grading, storage, drying, and monitoring. This section discusses artificial intelligence-based food monitoring and quality control systems. Numerous systems based on fuzzy logic were created.

13.14 Choosing Ideal Areas for Sowing Particular Crop:

AI may assist farm workers in selecting the best locations for crop planting based on the area's topography, the chemical composition of the soil, or any other factor by analyzing images from drones. The current crops will be identified, and the seed quality will be determined by a supervised machine learning algorithm. Before planting, AI might look at seed photos and compare them to images of seeds that are healthy. AI has effectively examined or monitored possible defects and nutrient shortages in the soil. The use of AI deep learning in agriculture supports the expansion of flora pattern analysis.

13.15 Weed Identification:

AI may be utilized to create image recognition software that has a high degree of accuracy in weed identification. Farmers and producers can use this to swiftly recognize and categorize various weed species and then implement the necessary control methods.

REFERENCES:

Benhmad, T., Rhaimi, C. B., Alomari, S., & Aljuhani, L. (2024). Design and Implementation of an Integrated IoT and Artificial Intelligence System for Smart Irrigation Management. *International Journal of Advances in Soft Computing & Its Applications*, *16*(1).

Bhati, R., & Mittal, S. (2023). The Role and Impact of Artificial Intelligence in Attaining Sustainability Goals. In *2023 9th International Conference on Advanced Computing and Communication Systems (ICACCS)*. DOI: 10.1109/ICACCS57279.2023.10112887

Chen, C.-J., Huang, Y.-Y., Li, Y.-S., Chang, C.-Y., & Huang, Y.-M. (2020). An AIoT Based Smart Agricultural System for Pests Detection. *IEEE Access : Practical Innovations, Open Solutions*, *8*, 180750–180761. DOI: 10.1109/ACCESS.2020.3024891

Eli-Chukwu, N. C. (2019). Applications of artificial intelligence in agriculture: A review. *Engineering, Technology & Applied Science Research*, *9*(4), 4. DOI: 10.48084/etasr.2756

Ennouri, K., Smaoui, S., Gharbi, Y., Cheffi, M., Ben Braiek, O., Ennouri, M., & Triki, M. A. (2021). Usage of artificial intelligence and remote sensing as efficient devices to increase agricultural system yields. *Journal of Food Quality*, *2021*(1), 6242288. DOI: 10.1155/2021/6242288

Hinnell, A. C., Lazarovitch, N., Furman, A., Poulton, M., & Warrick, A. W. (2010). Neuro-drip: Estimation of subsurface wetting patterns for drip irrigation using neural networks. *Irrigation Science*, *28*(6), 535–544. DOI: 10.1007/s00271-010-0214-8

Imalka, L. A., Gunawardana, K. G. A., Kodithuwakku, K. M. S. K., Arachchi, H. K. E., Harshanath, S. M. B., & Rajapaksha, S. (2022). Farming Through Technology Driven Solutions For Agriculture Industry Ceylon E-Agro mobile application-find technology based solutions for agricultural problems. In *2022 IEEE 10th Region 10 Humanitarian Technology Conference (R10-HTC)*. DOI: 10.1109/R10-HTC54060.2022.9929335

Javaid, M., Haleem, A., Khan, I. H., & Suman, R. (2023). Understanding the potential applications of Artificial Intelligence in Agriculture Sector. *Advanced Agrochem, 2*(1), 15-30. DOI: 10.1016/j.aac.2022.10.001

Jha, K., Doshi, A., Patel, P., & Shah, M. (2019). A comprehensive review on automation in agriculture using artificial intelligence. *Artificial Intelligence in Agriculture*, *2*, 1–12. DOI: 10.1016/j.aiia.2019.05.004

Kiani, F., Randazzo, G., Yelmen, I., Seyyedabbasi, A., Nematzadeh, S., Anka, F. A., Erenel, F., Zontul, M., Lanza, S., & Muzirafuti, A. (2022). A Smart and Mechanized Agricultural Application: From Cultivation to Harvest. *Applied Sciences (Basel, Switzerland)*, *12*(12), 6021. DOI: 10.3390/app12126021

Lochan, K., Khan, A., Elsayed, I., Suthar, B., Seneviratne, L., & Hussain, I. (2024). Advancements in Precision Spraying of Agricultural Robots: A Comprehensive Review. *IEEE Access : Practical Innovations, Open Solutions*, *12*, 129447–129483. DOI: 10.1109/ACCESS.2024.3450904

Mahlein, A.-K., Arnal Barbedo, J. G., Chiang, K.-S., Del Ponte, E. M., & Bock, C. H. (2024). From Detection to Protection: The Role of Optical Sensors, Robots, and Artificial Intelligence in Modern Plant Disease Management. *Phytopathology*, *114*(8), 1733–1741. DOI: 10.1094/PHYTO-01-24-0009-PER PMID: 38810274

Mahmud, M. S., Zahid, A., He, L., & Martin, P. (2021). Opportunities and possibilities of developing an advanced precision spraying system for tree fruits. *Sensors (Basel)*, *21*(9), 3262. DOI: 10.3390/s21093262 PMID: 34066785

Meenal, R., Binu, D., Ramya, K. C., Michael, P. A., Vinoth Kumar, K., Rajasekaran, E., & Sangeetha, B. (2022). Weather forecasting for renewable energy system: A review. *Archives of Computational Methods in Engineering*, *29*(5), 2875–2891. DOI: 10.1007/s11831-021-09695-3

. Miranda, J. L., Gerardo, B. D., & Tanguilig III, B. T. (2014). Pest detection and extraction using image processing techniques. International journal of computer and communication engineering, 3(3), 189-192.

Nyéki, A., & Neményi, M. (2022). Crop Yield Prediction in Precision Agriculture. *Agronomy (Basel), 12*(10), 2460. DOI: 10.3390/agronomy12102460

Ok, A. O., Akar, O., & Gungor, O. (2012). Evaluation of random forest method for agricultural crop classification. *European Journal of Remote Sensing*, *45*(1), 421–432. DOI: 10.5721/EuJRS20124535

. Preethi,. A. & Reddy, P. K. (2022). Survey on Smart Agriculture a Roadmap for India: Role of IoT, AI. *2022 3rd International Conference on Electronics and Sustainable Communication Systems (ICESC)*. .DOI: 10.1109/ICESC54411.2022.9885700

Pandian, P., Bose, P., Uma, M., Susithra, K., Panchami, B. G., & Reshma, R. (2024). Intelligent Sensor Based Smart Irrigation System For Sustainable Agriculture. *2024 International Conference on Communication, Computing and Internet of Things (IC3IoT)*. DOI: 10.1109/IC3IoT60841.2024.10550413

Panpatte, D. G. (2018). Artificial intelligence in agriculture: An emerging era of research. Anand Agricultural University, 1-8.

Pierre, N., Viviane, I. V. I., Lambert, U., Shadrack, I., Erneste, B., Schadrack, N., & Theogene, H. (2023). AI Based Real-Time Weather Condition Prediction with Optimized Agricultural Resources. *European Journal of Technology*, *7*(2), 36–49. DOI: 10.47672/ejt.1496

Premkumar, S., & Sigappi, A. N. (2022). IoT-enabled edge computing model for smart irrigation system. *Journal of Intelligent Systems*, *31*(1), 632–650. DOI: 10.1515/jisys-2022-0046

Qazi, S., Khawaja, B. A., & Farooq, Q. U. (2022). IoT-Equipped and AI-Enabled Next Generation Smart Agriculture: A Critical Review, Current Challenges and Future Trends. *IEEE Access : Practical Innovations, Open Solutions*, *10*, 21219–21235. DOI: 10.1109/ACCESS.2022.3152544

Rani, R., Sahoo, J., Bellamkonda, S., Kumar, S., & Pippal, S. K. (2023). Role of Artificial Intelligence in Agriculture: An Analysis and Advancements With Focus on Plant Diseases. *IEEE Access : Practical Innovations, Open Solutions*, *11*, 137999–138019. DOI: 10.1109/ACCESS.2023.3339375

Reddy, R., Sirish Kumar, M., Shereesha, M., Raju, K. S., Tangudu, N., & Nagendra Babu, G. (2023). A Comprehensive Overview on Implementing IoT-AI Interface in Agriculture. *2023 Second International Conference on Augmented Intelligence and Sustainable Systems (ICAISS)*. DOI: 10.1109/ICAISS58487.2023.10250697

Shaikh, F. K., Memon, M. A., Mahoto, N. A., Zeadally, S., & Nebhen, J. (2022). Artificial Intelligence Best Practices in Smart Agriculture. *IEEE Micro*, 42(1), 17-24. .DOI: 10.1109/MM.2021.3121279

. Shalini, V. T., Neware, R., Kumari, T., & Kumar, M. (2024). Soil Health Management Using Artificial Intelligence for Smart Agriculture Systems.

Singh, V., & Singh, G. (2023). Development of AI Enabled Solution for Efficient Feature Enrichment from Multiple Data Sources: An Application in Precision Agriculture. *2023 International Conference on Artificial Intelligence and Applications (ICAIA) Alliance Technology Conference (ATCON-1)*. DOI: 10.1109/ICAIA57370.2023.10169451

Sinwar, D., Dhaka, V. S., Sharma, M. K., & Rani, G. (2020). AI-based yield prediction and smart irrigation. *Internet of Things and Analytics for Agriculture*, *2*, 155–180.

Srivastava, P. K., Han, D., Rico-Ramirez, M. A., Bray, M., & Islam, T. (2012). Selection of classification techniques for land use/land cover change investigation. *Advances in Space Research*, *50*(9), 1250–1265. DOI: 10.1016/j.asr.2012.06.032

Subeesh, A., & Mehta, C. R. (2021). Automation and digitization of agriculture using artificial intelligence and internet of things. *Artificial Intelligence in Agriculture*, *5*, 278–291. DOI: 10.1016/j.aiia.2021.11.004

Ulfa, F., Orton, T. G., Dang, Y. P., & Menzies, N. W. (2022). Developing and Testing Remote-Sensing Indices to Represent within-Field Variation of Wheat Yields: Assessment of the Variation Explained by Simple Models. *Agronomy (Basel)*, *12*(2), 384. DOI: 10.3390/agronomy12020384

. Zha, J. (2020). Artificial intelligence in agriculture. *Journal of Physics: Conference Series*. IOP Publishing.

Chapter 3
Ethical Impacts of Using Artificial Intelligence in Sustainable Development: Challenges and Opportunities

Vishal Jain
https://orcid.org/0000-0003-1126-7424
Sharda University, India

Archan Mitra
https://orcid.org/0000-0002-1419-3558
NITTE University, India

ABSTRACT

The integration of artificial intelligence (AI) into sustainable development offers transformative opportunities for achieving the United Nations' Sustainable Development Goals (SDGs). AI technologies enhance efficiency in resource management, enable predictive analytics for climate action, and support inclusive economic growth and healthcare access. However, these advancements also introduce significant ethical challenges, including data privacy concerns, algorithmic biases, environmental impact, and accountability issues. This study explores the dual-edged impact of AI in

DOI: 10.4018/979-8-3693-8643-9.ch003

sustainable development, emphasizing the need for ethical frameworks to balance technological innovation with principles of equity, transparency, and sustainability. By examining case studies and existing policy frameworks, this research highlights both the benefits and risks of AI applications in sustainability and proposes actionable recommendations for ensuring responsible and equitable deployment.

I. INTRODUCTION

Contextual Background

The integration of artificial intelligence (AI) into sustainable development initiatives represents a dynamic frontier for addressing global challenges, aligning with the United Nations' Sustainable Development Goals (SDGs) established in 2015. The SDGs encompass 17 objectives designed to address critical issues such as poverty, hunger, clean water, climate change, and social inequalities, establishing a blueprint for achieving sustainable prosperity by 2030 (United Nations, 2015). Sustainable development aims to balance ecological, economic, and social priorities, creating an inclusive global framework that promotes prosperity without compromising future generations' ability to meet their needs. In this context, the SDGs emphasize the importance of resource management, energy efficiency, and environmental preservation, areas where AI can offer transformative solutions.

The rise of AI as a versatile and powerful technology has introduced advanced capabilities across sectors, ranging from healthcare to agriculture, transportation, and energy. AI-driven innovations can enable sustainable development by optimizing resource usage, enhancing predictive capabilities, and providing automated monitoring and real-time decision-making tools (Vinuesa et al., 2020). For instance, in agriculture, AI-based systems can predict weather patterns, optimize irrigation, and monitor crop health, thus maximizing productivity while reducing resource consumption (Sustainable Development Solutions Network, 2019). Similarly, in energy management, AI algorithms can manage electricity grids more efficiently, balancing demand and supply to reduce waste and reliance on non-renewable sources. However, while AI has the potential to catalyze the achievement of the SDGs, its application raises complex ethical issues that necessitate scrutiny.

Significance of the Study

The use of AI in sustainable development is marked by a dual-edged impact, as it enables substantial advancements but also introduces ethical challenges that cannot be ignored. AI applications can lead to unintended consequences such as data privacy infringements, algorithmic biases, and environmental impacts due to high computational requirements (Jobin et al., 2019). Moreover, AI-driven processes may exacerbate inequalities if access to technology remains concentrated among affluent groups or if AI applications fail to account for diverse socio-economic contexts. Therefore, understanding the ethical dimensions of AI's role in sustainable development is crucial for ensuring that AI not only drives innovation but does so responsibly and inclusively.

The significance of this study lies in its focus on the ethical dimensions of AI's application in sustainable development, examining both the benefits and challenges of implementing AI-driven solutions for the SDGs. This study will contribute to the academic and policy discourse by providing insights into how AI applications can be optimized to support sustainable development while adhering to ethical principles. Given the rapidly evolving nature of AI technologies, a comprehensive ethical assessment is essential for guiding future innovations and establishing trust among stakeholders, including governments, organizations, and communities.

Objectives

This research aims to explore the ethical impacts of using AI in sustainable development, seeking to identify both the challenges and opportunities associated with this technological advancement. By examining specific cases and frameworks, this study will provide a critical analysis of how AI applications can be both beneficial and potentially problematic when applied to sustainability efforts. The primary objectives include:

1. **Exploring the Ethical Dimensions of AI in Sustainable Development**

This study will examine how AI applications align with ethical considerations, focusing on principles such as fairness, accountability, transparency, and inclusivity. By identifying areas where AI's ethical implications are most pronounced, the study aims to highlight the potential risks and benefits of AI-driven sustainable solutions.

2. **Identifying Key Challenges and Opportunities in AI-Driven Sustainability**

This objective involves assessing the dual nature of AI in sustainability. While AI can enhance efficiencies and enable predictive capabilities for sustainable resource management, it may also introduce biases, compromise privacy, or unintentionally exacerbate socio-economic inequalities. The study will explore both the positive and negative impacts of AI, providing a balanced perspective on its role in achieving the SDGs.

3. **Recommending Ethical Frameworks and Policies for AI in Sustainability**

By examining current ethical frameworks and policies, this study will propose guidelines to ensure AI applications align with sustainability principles. These recommendations will address issues such as equitable access, data privacy, environmental impact, and algorithmic transparency, aiming to guide policymakers and developers in creating responsible and sustainable AI-driven solutions.

Research Questions

The following research questions guide this study, shaping its focus on the ethical dimensions of AI in sustainable development and facilitating a comprehensive examination of the associated challenges and opportunities:

1. **What Ethical Concerns Arise in the Use of AI for Sustainable Development?**

This question seeks to identify the ethical issues associated with using AI to support the SDGs. Topics of interest include data privacy, algorithmic fairness, environmental costs, and the potential for bias in AI-driven decision-making processes. This research question will guide the study's examination of specific ethical concerns, enabling a deeper understanding of AI's implications for sustainability.

2. **How Can These Challenges Be Mitigated to Harness AI's Potential Responsibly?**

Building on the ethical concerns identified, this question will explore potential strategies for mitigating negative impacts while maximizing AI's contributions to sustainable development. Solutions may include developing regulatory frameworks, implementing best practices for data management, and promoting transparency in AI algorithms.

3. **What Frameworks or Policies Are Necessary to Ensure Ethical AI Applications in Sustainability?**

This question addresses the need for comprehensive ethical frameworks and policies that can guide the responsible use of AI in sustainable development. The study will evaluate existing guidelines and propose new recommendations, aiming to establish a foundation for ethical AI that promotes inclusivity, accountability, and environmental stewardship.

The rapid advancement of AI presents a unique opportunity to address critical sustainability challenges, from optimizing resource management to enhancing predictive capabilities for climate action. However, the ethical dimensions of AI in sustainable development cannot be overlooked. This research seeks to contribute to the growing discourse on AI ethics by examining both the opportunities and challenges posed by AI applications in sustainability contexts. By exploring ethical concerns, mitigation strategies, and policy recommendations, this study aims to provide a comprehensive understanding of how AI can be leveraged responsibly in the pursuit of sustainable development. In doing so, it underscores the importance of aligning technological innovation with ethical principles, ensuring that AI-driven solutions contribute to a sustainable and equitable future for all.

II. LITERATURE REVIEW

AI in Sustainable Development

Artificial intelligence (AI) has rapidly emerged as a transformative force in sustainable development, with applications across various sectors, including agriculture, energy, healthcare, and climate science. Recent literature underscores AI's role in optimizing resource use, improving predictive capabilities, and providing data-driven insights that support decision-making in sustainability-related fields. For instance, AI-powered systems are being deployed in agriculture to enhance crop yield predictions, improve soil management, and optimize irrigation systems (Kyriazis et al., 2019). Machine learning algorithms, a subset of AI, analyze large datasets on climate patterns, soil composition, and crop health, enabling precise interventions that conserve resources while increasing productivity. In this way, AI directly contributes to Sustainable Development Goal (SDG) 2, which aims to achieve food security and promote sustainable agriculture (Vinuesa et al., 2020).

In the energy sector, AI applications are particularly relevant for managing renewable energy sources and enhancing the efficiency of energy systems. AI algorithms can balance electricity grids by predicting demand, optimizing energy storage, and facilitating the integration of renewable sources such as solar and wind power, thereby supporting SDG 7, which promotes affordable and clean energy for all (Sun et al., 2021). In healthcare, AI is being leveraged to monitor disease patterns, predict outbreaks, and improve diagnostic accuracy, contributing to SDG 3 by promoting good health and well-being (Topol, 2019). For instance, predictive modeling techniques aid in tracking infectious diseases and monitoring patients in remote regions, addressing healthcare access disparities.

Climate science also benefits from AI, as machine learning algorithms analyze complex environmental data to predict weather events, monitor deforestation, and model climate change scenarios. By enabling proactive climate action, AI supports SDG 13, which focuses on climate action (Rolnick et al., 2019). AI applications have also advanced sustainable urban planning (SDG 11), clean water management (SDG 6), and ecosystem conservation (SDG 15). Despite these positive contributions, the literature suggests that the ethical implications of AI's integration in sustainable development warrant thorough examination, as they are often complex and context-specific.

Ethical Frameworks and Principles for AI

As AI technologies continue to permeate sustainability initiatives, establishing ethical guidelines to govern their application is essential. Transparency, accountability, and fairness have been widely recognized as foundational principles in the ethical deployment of AI. These principles ensure that AI algorithms are explainable, that responsibility is assigned for AI-driven decisions, and that bias is minimized in decision-making processes (Jobin et al., 2019). The literature identifies transparency as a critical factor in fostering trust among stakeholders, as it allows users to understand how AI systems process data and reach conclusions (Floridi et al., 2018). Similarly, accountability is essential for ensuring that decision-makers remain responsible for AI outcomes, particularly when these decisions impact vulnerable populations.

Various organizations have established ethical frameworks to guide AI use. The IEEE's *Ethically Aligned Design* framework promotes principles such as human rights, well-being, and privacy, advocating for AI systems that support human values (IEEE, 2019). This framework emphasizes user-centered design, recommending that AI systems be tailored to serve users equitably and transparently. UNESCO's AI Ethics Recommendation extends these principles by focusing on inclusivity, environmental responsibility, and cultural diversity (UNESCO, 2021). The UNESCO guidelines urge developers to consider AI's social impact, emphasizing that AI technologies should prioritize public welfare and contribute positively to social equity. These ethical frameworks are integral to aligning

AI applications with sustainable development, as they provide a structured approach to addressing the ethical challenges associated with AI-driven sustainability initiatives.

Ethical principles like fairness and inclusivity are particularly relevant in sustainable development, where AI applications impact diverse socio-economic groups. Fairness in AI ensures that algorithms do not propagate biases against marginalized communities, which is crucial when AI applications influence resource allocation or access to essential services (West et al., 2019). Inclusivity is equally important, as it promotes equitable access to AI technologies and ensures that benefits are distributed fairly across all communities. These frameworks collectively create a foundation for ethical AI, highlighting the need for AI systems that are not only efficient but also aligned with the values of social and environmental justice.

Intersection of AI and Ethical Challenges in Sustainability

AI's application in sustainable development poses unique ethical challenges that are often interconnected with broader sustainability concerns. For instance, environmental justice is a critical issue in AI-driven projects that impact natural resources or public services. Ethical concerns arise when AI algorithms used in resource management inadvertently prioritize certain communities over others, leading to inequitable access to clean water, energy, or food resources (Binns, 2018). This issue is further complicated by the limited transparency in many AI systems, as algorithmic decision-making processes are often opaque, making it difficult for communities to understand or challenge resource allocations (Floridi et al., 2018).

Algorithmic bias is another pressing ethical challenge, as it can perpetuate social inequalities in sustainable development. For example, AI algorithms trained on biased datasets may disadvantage low-income or marginalized populations, especially in domains like healthcare and climate adaptation, where resource distribution is critical (Barocas et al., 2019). Data privacy is also a significant ethical concern, particularly in sustainability projects that rely on large-scale data collection for predictive analytics. In health monitoring or environmental tracking, AI systems often gather sensitive personal or ecological data, raising concerns about surveillance, consent, and the potential misuse of information (Crawford & Joler, 2018).

Resource allocation presents an additional ethical dilemma, as AI applications often require extensive computational power, leading to high energy consumption and carbon emissions. While AI can drive sustainability in sectors like energy and agriculture, it paradoxically generates environmental costs that challenge the very goals it seeks to promote (Strubell et al., 2019). This ethical tension underscores the importance of sustainable AI development, where the ecological footprint of AI technologies is carefully managed to align with environmental objectives.

Addressing these ethical challenges requires robust governance frameworks and a commitment to responsible innovation. Ethical AI principles like transparency, accountability, and inclusivity are crucial in mitigating biases and promoting fair outcomes in sustainable development. Governance frameworks can include regulatory guidelines that enforce data privacy, mandate transparency in AI-driven decisions, and promote equitable resource allocation. By adhering to these principles, AI can serve as a tool for sustainable development that respects human rights, minimizes ecological impacts, and promotes social justice.

III. METHODOLOGY

The methodology section outlines the research design, data collection methods, and analysis techniques employed to explore the ethical impacts of artificial intelligence (AI) in sustainable development. This study adopts a qualitative approach to capture the nuanced ethical considerations in applying AI to sustainable development goals (SDGs), focusing on case studies, policy analysis, and thematic exploration.

Research Design

This research employs a qualitative methodology centered on a combination of case studies, policy analysis, and a secondary data review. Qualitative research is particularly well-suited for exploring complex ethical issues as it allows for an in-depth examination of subjective and contextual elements, making it possible to capture the intricacies of ethical concerns in AI-driven sustainable development (Creswell, 2014). Given the multifaceted ethical landscape, the case study approach allows for focused, real-world examples where AI has been applied to specific SDGs, providing concrete insights into both the opportunities and challenges inherent in these applications.

In addition to case studies, policy analysis offers a critical lens for understanding the regulatory and ethical guidelines that shape AI's role in sustainability. By reviewing policies from organizations and international bodies such as UNESCO, the IEEE, and the United Nations, this study will examine how existing ethical frameworks are operationalized within the context of sustainable development (Patton, 2015). Secondary data review complements these approaches by gathering a range of information on AI applications in various sectors, providing a broad perspective on the intersection of AI ethics and sustainability. Together, these methods provide a comprehensive and robust research design that allows for an integrative understanding of ethical issues and the role of policy in mediating these challenges.

Data Collection

The data collection process involves gathering information from case studies of AI applications in sustainable development projects and reviewing relevant policy documents from leading organizations focused on AI ethics and sustainable development. Each data source offers unique insights into the ethical dimensions of AI in sustainability.

1. Case Studies of AI Applications in Sustainable Development

The case study component of this research focuses on specific AI applications in areas such as agriculture, healthcare, energy, and environmental conservation, aligning with key SDGs. For example, AI's role in optimizing irrigation in agriculture, predicting disease patterns in healthcare, and managing renewable energy in urban settings will be examined.

These case studies provide tangible examples of AI's potential benefits for sustainability as well as the ethical dilemmas that arise. Information for the case studies will be sourced from academic publications, reports by sustainability-focused organizations, and industry case reports from technology companies involved in AI applications for sustainability.

2. Policy Documents on AI Ethics and Sustainable Development

This study also involves a review of policy documents and ethical frameworks developed by organizations such as the IEEE, UNESCO, and the United Nations, which have established guidelines for responsible AI usage. The IEEE's *Ethically Aligned Design* report, UNESCO's *AI Ethics Recommendation*, and the United Nations' sustainability goals will serve as primary documents for this analysis (IEEE, 2019; UNESCO, 2021). These documents offer structured ethical frameworks that outline principles such as transparency, fairness, accountability, and inclusivity. Reviewing these policy documents allows the study to contextualize case study findings within broader regulatory and ethical guidelines, highlighting the alignment—or misalignment—between policy intentions and practical applications.

3. Secondary Data Review

Secondary data on AI's role in sustainable development is collected from peer-reviewed journal articles, technical reports, and sustainability assessments. This data source enriches the research by providing broader insights into the general trends, challenges, and opportunities associated with AI in sustainability. Key sources include databases such as IEEE Xplore, SpringerLink, and Google Scholar, which provide access to academic literature on AI ethics, sustainability studies, and policy evaluations. This review will help establish a solid foundation of knowledge on the current state of AI applications, relevant ethical debates, and existing policy interventions, allowing for a comprehensive understanding of the field.

Analysis Techniques

The data collected from case studies, policy documents, and secondary sources will undergo two main types of analysis: thematic analysis and comparative analysis. These techniques are chosen to provide a systematic exploration of ethical concerns and to assess the effectiveness of ethical frameworks in guiding sustainable AI practices.

1. Thematic Analysis of Ethical Concerns and Responses

Thematic analysis is a qualitative technique that identifies and examines patterns or themes within data. This study uses thematic analysis to uncover recurring ethical concerns, challenges, and responses across different AI applications in sustainability. Thematic coding will be used to categorize ethical issues, such as data privacy, algorithmic bias, environmental justice, and equitable access to resources (Braun & Clarke, 2006). This approach will allow for a nuanced understanding of how ethical challenges manifest in AI-driven sustainable development projects. For instance, themes related to fairness may highlight biases in AI models used for resource allocation, while transparency themes might expose challenges in making AI decision-making processes understandable to diverse stakeholders. Identifying these themes will provide insight into the most pressing ethical concerns and the ways that developers and policymakers are responding to these challenges.

2. Comparative Analysis of Policy Documents and Ethical Frameworks

Comparative analysis will be applied to assess the applicability and effectiveness of different ethical frameworks and policy documents in promoting sustainable AI. This analysis involves comparing the ethical guidelines proposed by the IEEE, UNESCO, and other bodies with the practical realities observed in the case studies. By examining similarities and differences among these frameworks, this study aims to evaluate their comprehensiveness and relevance to sustainability. The comparative analysis will consider factors such as the emphasis on specific ethical principles (e.g., inclusivity vs. accountability), the flexibility of the frameworks in adapting to diverse AI applications, and the extent to which these policies address specific sustainability concerns (Mills et al., 2021). This analysis will highlight gaps or limitations in current ethical guidelines and suggest areas for improvement to better support sustainable and ethical AI applications.

Thematic and comparative analyses, when combined, offer a holistic approach to understanding the ethical landscape of AI in sustainable development. Thematic analysis focuses on identifying and examining ethical concerns across various case studies, providing a grounded view of the challenges and responses within the field. Meanwhile, comparative analysis situates these findings within the broader policy environment, assessing the adequacy of existing ethical guidelines in addressing the identified concerns. Together, these methods contribute to a comprehensive exploration of how ethical challenges in AI-driven sustainable development are being addressed and where further policy intervention may be needed.

IV. ETHICAL CHALLENGES OF AI IN SUSTAINABLE DEVELOPMENT

The deployment of artificial intelligence (AI) in sustainable development initiatives presents significant ethical challenges. These challenges revolve around data privacy and security, bias and fairness, transparency and accountability, resource allocation and environmental impact, and autonomy and control. Understanding these ethical concerns is crucial for leveraging AI responsibly to achieve sustainable development goals (SDGs).

Data Privacy and Security

AI systems often rely on vast amounts of data to function effectively, particularly in areas like environmental monitoring and public health. This dependence on big data raises significant privacy concerns. In environmental monitoring, AI collects data from sensors and satellites to track deforestation, pollution levels, and wildlife populations (Gebru et al., 2021).

In health monitoring, AI analyzes patient data to predict disease outbreaks and personalize treatments (Topol, 2019). While these applications offer substantial benefits, they also risk exposing sensitive information.

A notable case illustrating data privacy breaches is the use of AI in smart city projects. In Songdo, South Korea, sensors collect data on residents' energy usage, waste disposal, and transportation habits to optimize city operations (Shin & Park, 2019). However, insufficient data anonymization led to concerns that individuals' behaviors could be tracked without consent, infringing on privacy rights. Similarly, during the COVID-19 pandemic, some governments utilized AI-driven contact tracing apps that collected location and health data without adequate privacy safeguards, leading to public outcry and mistrust (Morley et al., 2020).

These examples highlight the ethical imperative to balance data collection with robust privacy protections. Ensuring data is anonymized, securely stored, and collected transparently with informed consent is essential. Additionally, implementing strict data governance policies can mitigate risks associated with unauthorized access and potential misuse of personal information.

Bias and Fairness

AI algorithms can inadvertently perpetuate or even exacerbate existing societal biases, particularly affecting marginalized communities. This occurs when AI systems are trained on datasets that reflect historical inequalities, leading to biased outcomes in resource distribution (Barocas et al., 2019). In sustainable development, such biases can manifest in areas like access to clean energy, healthcare, and education.

For example, an AI system designed to allocate renewable energy resources may prioritize regions with better infrastructure data, often urban and wealthier areas, neglecting rural or impoverished communities (Mohamed et al., 2020). In developing regions, where data scarcity is common, AI models may underperform or produce skewed results, further marginalizing vulnerable populations (Buolamwini & Gebru, 2018).

Ensuring fairness in AI models requires diverse and representative datasets, as well as algorithms designed to detect and correct biases. Techniques such as fairness-aware machine learning can help mitigate these issues (Hardt et al., 2016). Moreover, involving local stakeholders in the development and deployment of AI systems can enhance cultural sensitivity and appropriateness, promoting equitable outcomes.

Transparency and Accountability

The opacity of AI algorithms, often referred to as the "black box" problem, poses significant ethical challenges in sustainability initiatives. When AI systems make decisions without clear explanations, it becomes difficult for stakeholders to understand, trust, or contest those decisions (Pasquale, 2015). This lack of transparency can undermine the legitimacy of sustainability projects and hinder community acceptance.

In the context of environmental conservation, AI may be used to enforce regulations by predicting illegal activities like poaching or logging. If the AI incorrectly identifies innocent parties due to opaque algorithms, it can lead to wrongful accusations without recourse (Vinuesa et al., 2020). Accountability becomes problematic when it's unclear who is responsible for AI-driven decisions—the developers, the operators, or the AI itself.

To address these issues, adopting explainable AI (XAI) methodologies can enhance transparency by making AI decision-making processes more understandable (Adadi & Berrada, 2018). Establishing clear accountability frameworks that define responsibility and liability for AI outcomes is also crucial. Regulatory policies should mandate documentation and audits of AI systems to ensure they adhere to ethical standards.

Resource Allocation and Environmental Impact

AI technologies often require significant computational resources, leading to high energy consumption and environmental costs. Training large AI models can result in substantial carbon emissions, counterintuitive to the goals of sustainable development (Strubell et al., 2019). For instance, developing a single natural language processing model can emit as much carbon as five cars in their lifetimes.

This paradox presents an ethical dilemma: utilizing AI to promote sustainability while contributing to environmental degradation. Sustainable development projects must consider the life-cycle environmental impact of AI technologies. For example, deploying AI in renewable energy management should account for the carbon footprint of the AI system itself.

Mitigating these concerns involves optimizing AI models for energy efficiency, utilizing renewable energy sources for data centers, and embracing green AI practices that prioritize resource-conscious approaches (Schwartz et al., 2020). Encouraging transparency in reporting the environmental impact of AI systems can also drive accountability and innovation toward more sustainable solutions.

Autonomy and Control

AI's capacity for autonomous decision-making raises ethical questions about human oversight and the potential consequences for natural resources and community livelihoods. In critical areas like resource management and conservation, AI systems may make decisions that affect large populations or ecosystems without human intervention.

For example, autonomous AI drones used for reforestation efforts might select planting sites based on algorithmic efficiency without considering local ecological knowledge or land rights, potentially disrupting existing land use or cultural practices (de Sousa et al., 2021). In agriculture, AI-driven automation could displace labor without adequate provisions for workers' livelihoods, exacerbating economic inequalities (Rotz et al., 2019). Ensuring appropriate levels of human oversight is essential to address these ethical concerns. Implementing human-in-the-loop systems, where human judgment complements AI decision-making, can help balance efficiency with ethical considerations (Rahwan, 2018). Engaging local communities in the deployment of AI applications ensures that technologies align with societal values and do not infringe upon human rights or cultural norms.

The ethical challenges associated with AI in sustainable development are multifaceted and significant. Addressing data privacy and security requires robust safeguards and transparent data practices. Mitigating bias and ensuring fairness necessitate diverse datasets and inclusive algorithm design. Enhancing transparency and accountability involves adopting explainable AI and clear responsibility frameworks. Balancing AI's environmental impact with sustainability goals calls for energy-efficient models and green AI practices. Finally, maintaining autonomy and control requires human oversight and community engagement. Recognizing and proactively addressing these ethical challenges is crucial for the responsible integration of AI into sustainable development initiatives. By doing so, we can harness AI's potential to advance the SDGs while upholding ethical standards that protect individuals, communities, and the environment.

V. OPPORTUNITIES OF AI IN SUSTAINABLE DEVELOPMENT

Artificial intelligence (AI) presents transformative opportunities for sustainable development across various domains, including resource management, climate action, economic growth, and public health. By enhancing efficiency, facilitating predictive analytics, and promoting inclusive growth, AI has the potential to contribute significantly to achieving the Sustainable Development Goals (SDGs) established by the United Nations (UN). However, leveraging these opportunities requires balancing the benefits with ethical considerations to ensure that AI-driven advancements promote sustainability in a responsible and equitable manner.

Enhanced Efficiency and Resource Management

One of AI's most promising contributions to sustainable development lies in its ability to optimize resource use, thereby enhancing efficiency and reducing waste. Case studies from multiple sectors illustrate how AI-driven systems can support sustainable resource management. In water management, for instance, AI-based platforms analyze data from sensors, satellite imagery, and weather forecasts to improve irrigation practices, conserve water, and enhance crop yields (Sustainable Development Solutions Network, 2019). By adjusting irrigation based on real-time soil moisture levels and climate conditions, AI helps farmers avoid overwatering, thereby conserving water resources while maintaining agricultural productivity.

In energy conservation, AI algorithms are employed to monitor and manage electricity grids. AI systems can predict energy demand patterns, enabling the efficient allocation of renewable energy resources like wind and solar power. For example, DeepMind, Google's AI subsidiary, uses AI to predict and optimize energy needs at its data centers, resulting in a 40% reduction in cooling energy usage (Evans & Gao, 2016). This reduction in energy consumption not only lowers operational costs but also reduces carbon emissions, thus aligning with climate action goals.

Waste reduction is another area where AI demonstrates its potential. AI-driven sorting systems in recycling facilities can automatically identify and separate recyclable materials from waste, reducing the contamination of recycling streams and increasing the efficiency of waste processing (Kaza et al., 2018). Moreover, predictive algorithms in manufacturing can minimize waste by identifying inefficiencies in production lines and reducing material usage. The ethical advantage of AI in these applications lies in its ability to promote efficiency, reduce resource consumption, and support environmental sustainability, aligning closely with SDGs 6 (Clean Water and Sanitation), 7 (Affordable and Clean Energy), and 12 (Responsible Consumption and Production).

Predictive Analytics for Climate Action

AI's capabilities in predictive analytics offer valuable tools for climate action, particularly in modeling climate patterns, forecasting natural disasters, and guiding mitigation efforts. Predictive models powered by AI can process vast amounts of data to generate more accurate climate forecasts, enabling governments and organizations to make proactive decisions for climate resilience (Rolnick et al., 2019). For instance, AI models have been used to predict the spread of wildfires, allowing authorities to implement preventive measures and allocate resources more effectively. In California, for example, AI-driven predictive systems have been deployed to assess fire risk and warn communities, improving preparedness and minimizing loss of life and property (Harris & Johnson, 2020).

Beyond disaster prediction, AI's predictive analytics contribute to long-term climate mitigation strategies. Machine learning algorithms analyze historical data on greenhouse gas emissions, land use changes, and energy consumption to forecast future environmental impacts, guiding policymakers in setting realistic carbon reduction targets (Vinuesa et al., 2020). The ethical advantage of AI in climate action is its capacity to support evidence-based decisions, enhance community resilience, and improve preparedness for climate-related events, contributing to SDG 13 (Climate Action). However, these applications must ensure equitable access to climate information to prevent disparities in preparedness among different communities.

Inclusive Economic Growth

AI has the potential to foster inclusive economic growth by enabling automation and innovation across various sectors, thus creating job opportunities and reducing poverty. In the agricultural sector, AI-powered technologies such as automated harvesters and precision farming tools increase productivity, which can enhance food security and improve farmers' livelihoods. In developing countries, small-scale farmers benefit from AI applications that provide real-time insights on crop health, market prices, and climate forecasts, enabling them to make informed decisions and increase profitability (Chui et al., 2021).

In manufacturing, AI-driven automation optimizes production processes, reduces operational costs, and promotes economic growth by increasing the competitiveness of local industries. AI-driven systems in logistics and supply chain management improve efficiency, allowing businesses to reduce costs and pass these savings on to consumers, contributing to a more inclusive economy. However, the ethical consideration here revolves around ensuring that AI-driven automation does not displace low-skilled workers without adequate retraining programs. By investing in workforce reskilling initiatives, governments and businesses can facilitate a smoother transition for workers affected by automation, promoting equitable development and reducing poverty in line with SDG 8 (Decent Work and Economic Growth) and SDG 10 (Reduced Inequalities).

The ethical advantage of AI in inclusive economic growth lies in its potential to democratize access to information, enable productivity gains, and stimulate innovation. However, equitable implementation requires careful policy design to ensure that AI's benefits reach marginalized populations, especially in developing regions.

Advancement of Public Health

AI's contributions to public health present significant opportunities for promoting well-being and enhancing healthcare accessibility, particularly in underserved regions. AI-driven diagnostic tools and health monitoring systems improve the accuracy of medical diagnoses, enabling healthcare providers to detect diseases early and initiate treatment more effectively. In rural and remote areas, AI-powered telemedicine platforms facilitate access to healthcare services by connecting patients with specialists, reducing travel costs and improving health outcomes (Topol, 2019).

AI also plays a critical role in monitoring and predicting disease outbreaks. During the COVID-19 pandemic, AI models were used to analyze data from diverse sources, including social media, news reports, and health records, to track the spread of the virus and inform public health responses (Vaishya et al., 2020). By identifying early signals of disease outbreaks, AI enables timely intervention, reducing the spread of infectious diseases and potentially saving lives.

While AI's applications in public health offer numerous benefits, they also raise ethical challenges related to data privacy and security. The use of sensitive health data requires stringent safeguards to protect patient confidentiality and prevent misuse. To address these concerns, healthcare organizations must implement robust privacy policies and transparency practices, ensuring that AI-driven health systems comply with ethical standards and protect individual rights. The ethical benefits of AI in public health align with SDG 3 (Good Health and Well-being) by promoting access to quality healthcare and improving health equity.

The opportunities presented by AI in sustainable development are vast, with the potential to enhance resource efficiency, enable predictive climate action, foster inclusive economic growth, and advance public health. These applications demonstrate AI's capability to contribute positively to the SDGs by addressing critical issues such as water and energy conservation, climate resilience, economic empowerment, and healthcare access. However, to maximize these benefits, AI implementations must be guided by ethical principles that prioritize equity, transparency, and privacy. Balancing AI's technological advantages with ethical considerations will ensure that sustainable development initiatives not only achieve efficiency and growth but also promote fairness and inclusivity. By investing in equitable access, transparent practices, and data security, governments, businesses, and organizations can harness AI's potential to build a sustainable and just future. As AI continues to evolve, maintaining a commitment to ethical guidelines will be essential for driving positive change in alignment with global sustainability goals.

VI. DISCUSSION

Artificial intelligence (AI) presents a dual-edged impact on sustainable development, with the potential to accelerate progress toward the Sustainable Development Goals (SDGs) while simultaneously raising significant ethical challenges. This discussion integrates insights from the preceding sections, synthesizing the opportunities AI offers for sustainable development and the ethical challenges it introduces. The balance between leveraging AI's positive impact and addressing ethical concerns is critical to ensuring responsible AI deployment. In this context, policy frameworks become indispensable tools for setting boundaries and guiding AI's role in sustainable development.

Synthesis of Challenges and Opportunities

AI's impact on sustainable development is characterized by its ability to enhance efficiency, predictive analytics, economic inclusivity, and healthcare access. These opportunities directly support the SDGs, contributing to resource conservation, climate action, equitable growth, and improved health outcomes. For instance, AI-powered tools optimize resource management, helping reduce water and energy waste in agriculture and urban settings (Evans & Gao, 2016; Sustainable Development Solutions Network, 2019). Predictive models help governments and organizations anticipate climate events, thus enabling proactive measures to mitigate damage (Rolnick et al., 2019). Furthermore, AI-driven automation can support inclusive economic growth by enhancing productivity in agriculture and manufacturing, especially in developing regions (Chui et al., 2021).

Despite these promising benefits, AI applications are not without ethical risks. Privacy concerns arise due to AI's dependency on large datasets, which, if mismanaged, can lead to breaches of personal information (Morley et al., 2020). Bias and fairness issues are prevalent when AI systems inadvertently perpetuate social inequalities, affecting marginalized communities (Buolamwini & Gebru, 2018). Additionally, transparency and accountability are challenging to maintain due to the complexity of AI algorithms, which may lead to "black box" scenarios where decisions are made without clear explanations (Pasquale, 2015). AI's resource-intensive nature also poses environmental concerns, as the energy requirements of large AI models contribute to carbon emissions (Strubell et al., 2019).

The synthesis of these challenges and opportunities underscores the need for a nuanced approach to AI in sustainable development. While AI holds transformative potential, its deployment must be carefully managed to avoid exacerbating social and environmental issues. Understanding AI's dual-edged nature is essential for developing strategies that maximize its benefits while minimizing risks, thus supporting sustainable development in a responsible and ethical manner.

Balancing Ethical Concerns with Sustainable Goals

The balance between maximizing AI's positive impact on sustainable development and addressing ethical concerns is complex and multifaceted. On one hand, AI offers unprecedented opportunities for achieving sustainability by enhancing efficiency, promoting inclusivity, and supporting climate resilience. These applications can significantly contribute to the SDGs by driving advancements in clean energy, waste reduction, equitable growth, and public health (Vinuesa et al., 2020). However, ignoring ethical considerations in AI-driven sustainability can have severe implications, including the potential to deepen social divides, erode privacy rights, and harm the environment.

When ethical concerns are overlooked, AI applications in sustainable development risk perpetuating harm rather than fostering positive change. For instance, an AI system that prioritizes efficiency over fairness in resource allocation may unintentionally disadvantage low-income communities or reinforce existing inequalities (Mohamed et al., 2020). In healthcare, AI's capability to predict and monitor diseases can improve access to medical care, but if privacy protections are neglected, it could lead to the misuse of sensitive health data, eroding public trust (Topol, 2019). Additionally, without transparency and accountability, AI-driven decision-making in sustainability projects may lack legitimacy, making it difficult for affected communities to contest or understand these decisions (Pasquale, 2015).

Ethical concerns must therefore be addressed alongside AI's sustainability goals to ensure that the benefits of AI are shared equitably and responsibly. One way to achieve this balance is through the adoption of ethical principles, such as fairness, accountability, transparency, and privacy, in the development and deployment of AI systems. By embedding these principles in AI applications, developers and policymakers can create frameworks that protect individual rights while supporting sustainable development. The integration of ethical safeguards into AI-driven sustainability initiatives ensures that the pursuit of efficiency does not come at the expense of social and environmental justice.

Policy Implications

Given the complex interplay between AI's benefits and ethical risks, policy frameworks are essential for ensuring responsible AI deployment in sustainable development. Policies should aim to balance the need for innovation with ethical safeguards, thus promoting AI's role in sustainability while protecting individuals, communities, and the environment. Effective policy frameworks can establish guidelines for data privacy, fairness, accountability, and transparency in AI applications, aligning with the broader goals of sustainable development.

Recommendations for Policy Frameworks

1. **Data Privacy and Security Protections** Policies must prioritize data privacy by establishing standards for data collection, storage, and sharing in AI-driven sustainability projects. Regulatory frameworks, such as the General Data Protection Regulation (GDPR) in the European Union, offer a model for protecting personal information and could serve as a foundation for AI policies worldwide (European Parliament, 2016). These protections should ensure that data used in AI systems is anonymized, securely stored, and accessed only with informed consent.
2. **Algorithmic Fairness and Bias Mitigation** Policymakers should implement fairness standards in AI development, requiring developers to use diverse and representative datasets. Fairness-aware algorithms can help identify and mitigate biases, ensuring that AI systems do not discriminate against marginalized communities (Hardt et al., 2016). Additionally, policies can mandate the auditing of AI models for biases and require transparency in the decision-making criteria of AI systems.
3. **Transparency and Accountability Standards** Policies that encourage explainable AI (XAI) techniques are crucial for enhancing transparency in AI-driven decisions. Clear guidelines should define accountability for AI-related outcomes, especially in areas impacting public resources and safety (Adadi & Berrada, 2018). Policymakers can require AI developers to document decision-making processes, enabling stakeholders to understand, contest, and correct AI-driven outcomes.
4. **Environmental Impact Considerations** AI policies should address the environmental footprint of AI models by encouraging the use of energy-efficient models and renewable energy sources in data centers. This "green AI" approach can help reduce the carbon emissions associated with AI, supporting climate goals while promoting technological innovation (Schwartz et al., 2020). Policymakers might also incentivize research into low-power AI models that minimize resource consumption.

The Role of Governments, Corporations, and International Organizations

The responsibility for setting guidelines for AI in sustainability does not rest with any single entity; it requires collaboration among governments, corporations, and international organizations. Governments can play a regulatory role by enacting laws that protect public interests and hold developers accountable for ethical AI use. Corporations, particularly those involved in AI development and deployment, have a responsibility to adhere to ethical guidelines and implement robust internal policies that promote transparency, fairness, and accountability.

International organizations, such as the United Nations and the Organisation for Economic Co-operation and Development (OECD), can facilitate global cooperation by developing universal AI ethics standards. The United Nations, for instance, has already emphasized the role of technology in achieving the SDGs, making it a suitable platform for coordinating AI ethics in sustainable development (UNESCO, 2021). These organizations can foster cross-border partnerships, promote knowledge exchange, and establish consistent standards that transcend national boundaries, ensuring that AI's role in sustainability is guided by a unified ethical framework.

AI's application in sustainable development offers immense potential, yet it presents complex ethical challenges that must be addressed to ensure responsible and equitable outcomes. The synthesis of these challenges and opportunities highlights the need for a balanced approach that maximizes AI's contributions to sustainability while safeguarding ethical principles. Policymakers, governments, corporations, and international organizations all have a critical role to play in establishing policies that balance innovation with ethical safeguards. As AI continues to evolve, its integration into sustainable development requires constant vigilance and a commitment to ethical principles. By adopting comprehensive policy frameworks and fostering international collaboration, society can harness AI's potential to support the SDGs in a manner that promotes social equity, protects privacy, and minimizes environmental impact. Through these efforts, AI can become a powerful tool for sustainable development, contributing to a just and resilient future.

VII. CONCLUSION

Artificial intelligence (AI) stands as a transformative force with the potential to drive sustainable development across critical areas such as resource management, climate resilience, economic growth, and public health. The insights gathered in this research highlight AI's dual-edged impact in achieving the Sustainable Development Goals (SDGs): on the one hand, AI can greatly enhance efficiency, enable data-driven decision-making, and support inclusive growth, contributing positively to environmental sustainability and social equity. On the other hand, AI raises substantial ethical challenges, including issues related to data privacy, bias, transparency, resource consumption, and autonomy. These challenges must be carefully managed to ensure AI's deployment aligns with the ethical standards essential for fostering a just and sustainable future.

The discussion has shown that balancing AI's technological advancements with ethical considerations is paramount to achieving responsible AI in sustainable development. Failing to address these ethical issues risks unintended consequences, such as perpetuating inequality, compromising privacy, undermining accountability, and increasing environmental burdens. Thus, this balance is crucial not only for the integrity of AI-driven initiatives but also for the broader goals of social and environmental justice embedded in the SDGs.

The role of policy frameworks emerges as critical in this balancing act. Policies designed to protect data privacy, promote fairness, ensure transparency, and mitigate environmental impact can serve as the foundation for responsible AI use. Governments, corporations, and international organizations all play essential roles in this endeavor, with each stakeholder contributing to the establishment of guidelines and standards that ensure AI serves humanity's best interests. These entities must work collaboratively to create a cohesive, global approach to ethical AI in sustainability, transcending national boundaries and fostering shared responsibility.

In conclusion, AI offers substantial opportunities to support sustainable development, but its ethical challenges demand rigorous oversight, thoughtful policies, and ongoing collaboration. By embedding ethical principles into AI's development and deployment, society can harness

AI's potential to not only accelerate progress toward the SDGs but to do so in a way that is just, equitable, and sustainable. Achieving this vision requires a commitment to responsible innovation, ethical governance, and continuous engagement with communities, ensuring that AI remains a tool for positive change in both present and future generations.

REFERENCES

Adadi, A., & Berrada, M. (2018). Peeking inside the black-box: A survey on explainable artificial intelligence (XAI). *IEEE Access : Practical Innovations, Open Solutions*, *6*, 52138–52160. DOI: 10.1109/ACCESS.2018.2870052

Barocas, S., Hardt, M., & Narayanan, A. (2019). *Fairness and machine learning*. http://fairmlbook.org

Binns, R. (2018). Fairness in machine learning: Lessons from political philosophy. *Proceedings of the 2018 Conference on Fairness, Accountability, and Transparency*, 149-159. DOI: 10.1145/3287560.3287576

Braun, V., & Clarke, V. (2006). Using thematic analysis in psychology. *Qualitative Research in Psychology*, *3*(2), 77–101. DOI: 10.1191/1478088706qp063oa

Buolamwini, J., & Gebru, T. (2018). Gender shades: Intersectional accuracy disparities in commercial gender classification. *Proceedings of Machine Learning Research*, *81*, 1–15.

Chui, M., Manyika, J., & Miremadi, M. (2021). *The impact of AI on work: An economy-wide view*. McKinsey Global Institute.

Crawford, K., & Joler, V. (2018). Anatomy of an AI system: The Amazon Echo as an anatomical map of human labor, data, and planetary resources. *AI Now Institute*. https://anatomyof.ai

Creswell, J. W. (2014). *Research design: Qualitative, quantitative, and mixed methods approaches* (4th ed.). SAGE Publications.

de Sousa, W. G., Khosravi, S., Cristoni, R., & Ketelhodt, L. (2021). The role of AI-powered drones in sustainable development: Evidence from reforestation projects. *Sustainability*, *13*(15), 8505. DOI: 10.3390/su13158505

European Parliament. (2016). *Regulation (EU) 2016/679 of the European Parliament and of the Council of 27 April 2016*. General Data Protection Regulation.

Evans, R., & Gao, J. (2016). *DeepMind AI reduces Google data centre cooling bill by 40*. DeepMind.
Evans, R., & Gao, J. (2016). *DeepMind AI reduces Google data centre cooling bill by 40*. DeepMind.
Floridi, L., Cowls, J., Beltrametti, M., Chatila, R., Chazerand, P., Dignum, V., Luetge, C., Madelin, R., Pagallo, U., Rossi, F., Schafer, B., Valcke, P., & Vayena, E. (2018). AI4People—An ethical framework for a good AI society: Opportunities, risks, principles, and recommendations. *Minds and Machines*, *28*(4), 689–707. DOI: 10.1007/s11023-018-9482-5 PMID: 30930541

Gebru, T., Morgenstern, J., Vecchione, B., Vaughan, J. W., Wallach, H., Daumé, H.III, & Crawford, K. (2021). Datasheets for datasets. *Communications of the ACM*, *64*(12), 86–92. DOI: 10.1145/3458723

Hardt, M., Price, E., & Srebro, N. (2016). Equality of opportunity in supervised learning. *Advances in Neural Information Processing Systems, 29*, 3315–3323. https://proceedings.neurips.cc/paper/2016/hash/a486cd07e4ac3d270571622f4f316ec5-Abstract.html

Harris, M., & Johnson, C. (2020). Using AI to predict wildfires and mitigate their effects: A case study from California. *Journal of Climate Resilience*, *7*(3), 215–228.

IEEE. (2019). *Ethically Aligned Design: A Vision for Prioritizing Human Well-being with Autonomous and Intelligent Systems*. https://standards.ieee.org/industry-connections/ec/ead.html
IEEE. (2019). *Ethically Aligned Design: A Vision for Prioritizing Human Well-being with Autonomous and Intelligent Systems*. https://standards.ieee.org/industry-connections/ec/ead.html
Jobin, A., Ienca, M., & Vayena, E. (2019). The global landscape of AI ethics guidelines. *Nature Machine Intelligence*, *1*(9), 389–399. DOI: 10.1038/s42256-019-0088-2

Kaza, S., Yao, L. C., Bhada-Tata, P., & Van Woerden, F. (2018). *What a waste 2.0: A global snapshot of solid waste management to 2050*. World Bank Publications. DOI: 10.1596/978-1-4648-1329-0

Kyriazis, D., Varvarigou, T., White, D., & Rossi, A. (2019). Sustainable smart city IoT applications: A novel eco-social model. *Internet of Things : Engineering Cyber Physical Human Systems, 1-2*, 4–10. DOI: 10.1016/j.iot.2018.08.001

Mills, A. J., Durepos, G., & Wiebe, E. (Eds.). (2021). *Encyclopedia of Case Study Research*. SAGE Publications.

Mohamed, S., Png, M.-T., & Isaac, W. (2020). Decolonial AI: Decolonial theory as sociotechnical foresight in artificial intelligence. *Philosophy & Technology, 33*(4), 659–684. DOI: 10.1007/s13347-020-00405-8

Mohamed, S., Png, M.-T., & Isaac, W. (2020). Decolonial AI: Decolonial theory as sociotechnical foresight in artificial intelligence. *Philosophy & Technology, 33*(4), 659–684. DOI: 10.1007/s13347-020-00405-8

Morley, J., Cowls, J., Taddeo, M., & Floridi, L. (2020). Ethical guidelines for COVID-19 tracing apps. *Nature, 582*(7810), 29–31. DOI: 10.1038/d41586-020-01578-0 PMID: 32467596

Pasquale, F. (2015). *The black box society: The secret algorithms that control money and information*. Harvard University Press. DOI: 10.4159/harvard.9780674736061

Pasquale, F. (2015). *The black box society: The secret algorithms that control money and information*. Harvard University Press. DOI: 10.4159/harvard.9780674736061

Patton, M. Q. (2015). *Qualitative research & evaluation methods: Integrating theory and practice* (4th ed.). SAGE Publications.

Rahwan, I. (2018). Society-in-the-loop: Programming the algorithmic social contract. *Ethics and Information Technology, 20*(1), 5–14. DOI: 10.1007/s10676-017-9430-8

Rolnick, D., Donti, P. L., Kaack, L. H., Kochanski, K., Lacoste, A., Sankaran, K., . . . Bengio, Y. (2019). Tackling climate change with machine learning. *arXiv preprint arXiv:1906.05433*. https://arxiv.org/abs/1906.05433

Rotz, S., Gravely, E., Mosby, I., Duncan, E., Finnis, E., Horgan, M., LeBlanc, J., Martin, R., Neufeld, H. T., Nixon, A., Pant, L., Shalla, V., & Fraser, E. D. G. (2019). Automated pastures and the digital divide: How agricultural technologies are shaping labour and rural communities. *Journal of Rural Studies*, *68*, 112–122. DOI: 10.1016/j.jrurstud.2019.01.023

Schwartz, R., Dodge, J., Smith, N., & Etzioni, O. (2020). Green AI. *Communications of the ACM*, *63*(12), 54–63. DOI: 10.1145/3381831

Shin, D.-H., & Park, Y. J. (2019). Role of fairness, accountability, and transparency in algorithmic affordance. *Computers in Human Behavior*, *98*, 277–284. DOI: 10.1016/j.chb.2019.04.019

Strubell, E., Ganesh, A., & McCallum, A. (2019). Energy and policy considerations for deep learning in NLP. *Proceedings of the 57th Annual Meeting of the Association for Computational Linguistics*, 3645–3650. DOI: 10.18653/v1/P19-1355

Sun, Y., Song, H., Jara, A. J., & Bie, R. (2021). Internet of Things and big data analytics for smart and connected communities. *Journal of Environmental Research and Public Health*, *18*(9), 4784. DOI: 10.3390/ijerph18094784 PMID: 33946141

Sustainable Development Solutions Network. (2019). *The role of technology in delivering the sustainable development goals*. United Nations.

Topol, E. J. (2019). *Deep medicine: How artificial intelligence can make healthcare human again*. Basic Books.

UNESCO. (2021). *Recommendation on the Ethics of Artificial Intelligence*. United Nations Educational, Scientific and Cultural Organization.

UNESCO. (2021). *Recommendation on the Ethics of Artificial Intelligence*. https://unesdoc.unesco.org

United Nations. (2015). *Transforming our world: The 2030 agenda for sustainable development*.

Vaishya, R., Javaid, M., Khan, I. H., & Haleem, A. (2020). Artificial intelligence (AI) applications for COVID-19 pandemic. *Diabetes & Metabolic Syndrome*, *14*(4), 337–339. DOI: 10.1016/j.dsx.2020.04.012 PMID: 32305024

Vinuesa, R., Azizpour, H., Leite, I., Balaam, M., Dignum, V., Domisch, S., & Nerini, F. F. (2020). The role of artificial intelligence in achieving the Sustainable Development Goals. *Nature Communications*, *11*(1), 233. DOI: 10.1038/s41467-019-14108-y PMID: 31932590

Chapter 4

Zero Trust Architecture-Based Enhanced Security Framework for OTT in Hybrid Cloud Environment

Biswaranjan Jena
Siksha 'O' Anusandhan, India

Debahuti Mishra
Siksha 'O' Anusandhan, India

Smitaprava Mishra
https://orcid.org/0000-0001-9166-829X
Siksha 'O' Anusandhan, India

Sashikala Mishra
https://orcid.org/0000-0002-5433-4917
Symbiosis Institute of Technology, India

ABSTRACT

Growth and availability of internet with significant higher internet bandwidth and lower cost has fuelled the growth of Over the Top (OTT) platforms. Multiple channels like mobile, tv, set top boxes including gaming platforms need to access the content in secure way. As most of the revenue for an OTT platform comes from subscription and ads, it's crucial to show the right content to authorized users while maintaining the privacy of users following General Data Protection Regulation (GDPR) guide-

DOI: 10.4018/979-8-3693-8643-9.ch004

lines. This chapter proposes a security architecture framework for OTT which can be used to achieve higher level of security following zero-trust architecture principles. Each component of the framework from frontend to middleware to backend along with cloud infrastructure is explained in detail with best practices followed for achieving security.

1. INTRODUCTION

Over The Top (OTT) platforms are the recent trend in the media and entertainment industry where millions of viewers can view their favorite contents anytime, anywhere on any device. As per Statista (2022) the Indian OTT market revenue is expected to grow to $4144 million with user penetration of 31.6% by 2024. With millions of users watching on open internet, security of content and services exposed as APIs to end user devices is of paramount importance. Any breach can lead to loss of market reputation and loss of revenue. This chapter briefly describes various services and how security could be taken care at the middleware both from infrastructure and application level, to provide secure APIs to be consumed from anywhere by any channel. The principles applied here can be extended to any other domain too where the business logic is running on containers and exposed as APIs to external world. Gone are those days where services are running on legacy monolithic servers which has its own challenges of provisioning, maintenance, cost overhead and scalability. Enterprises are now relying on container strategies using Docker, Kubernetes, OpenShift type of solutions which are not only ready to provision and scale faster but also can be chosen as managed services as provided by various cloud providers like Amazon Web Services (AWS), Google, Azure and many more. This chapter explains zero trust architecture principles and conceptual framework to achieve highest level of security for OTT platforms starting from end-user device to middleware. Infrastructure security is explained with reference to AWS cloud although the concepts can be extended to other cloud infrastructures like Google and Microsoft Azure. This chapter clearly depicts the principles and methodology adopted at various layers like infrastructure, application, containerization, network, data storage etc. keeping in view of OTT reference architecture.

02. LITERATURE STUDY

Technology has moved ahead, in one hand with more tools and frameworks coming up for enhanced security in other hand hackers find out ways to exploit them. Gone are those days of perimeter security where the systems were completely within corporate firewalls and IP/Port restrictions and enterprise assumed their systems were secure. To overcome this John Kidervag of Forrester Research has coined Zero Trust (Goodman, 2019). It encourages frequent evaluation related to resource protection and discourages implicit grant to resources as described by Scott Rose et. al. (n.d.). This encourages access to resource via proper authentication and authorization only. With evolution of cloud infrastructure and applications are accessible from anywhere via any device the threat surface has further increased with average of five breaches per enterprise as published by Praful Pandey et. al. (2019). Daniel D'Silva et. al. (2021) have described how to achieve zero trust architecture in Kubernetes environment. In most of the enterprises now, including OTT platforms, services are deployed in some or the other container platform and the principles outlined here help them to achieve higher security goals. The overall principle of zero trust architecture revolved around "Trust" which assumes that no component or resource within or outside specific infrastructure is trustable and hence each access points should be authenticated and verified before giving access to resource. This also includes principles like frequent monitoring and re-evaluation too. Baozhan Chen et. al. (2021) proposes a security awareness and protection system for a 5G based smart healthcare system. Considerable threat exists to personal Medicare data which used to be held centrally during initial days, but with 5G network it becomes online and can be accessed outside traditional boundary. Zero Trust Architecture (ZTA) treats any person, event or device inside network and outside network as not trustworthy. These resources are trusted only after sufficient verification and authentication. The proposed framework focuses on fine-grained access control, authenticating identity, following least privilege principle, and analyzing the access pattern. Wu Qiang et. al. (2021) proposed intrinsic cloud security, a novel paradigm which provides dynamic defense to cloud native environment which provides much improved performance and cost reduction while reducing attack surface.

In this article we have offered following contributions:

- Highlighted various security threats that can be faced by OTT system.
- Proposed security framework that can be implemented by any OTT.
- Described how security can be controlled across components of the framework.
- Describe how to identify any security threat using various tools and appropriate control measures to be taken to protect data and system.

The remainder of the article is organized as follows. Section 03 briefly describes the OTT architecture and various components or systems who play vital role in providing the service. Amazon Web Service (AWS) is chosen as the cloud platform to provide the required infrastructure, but the same concept can be implemented using other tools as available in other cloud providers like Google GCP and Microsoft Azure.

03. OTT ARCHITECTURE AND COMPONENTS

Security is never achieved in isolation and each component or layer as shown in **Figure 1** plays role to achieve the highest level of security. Data is encapsulated across multiple layers to provide the best protection, a concept called defense in depth. Infrastructure security is used to protect data center devices like racks, servers, routers etc. Various components like firewall, load balancer, VPN, intrusion detection system, network access control etc plays critical role in providing security at the network layer. While cloud infrastructure provide physical security to their data centers as binded by various Service Level Agreements (SLA), data center security is managed by organization by enforcing various policies like CCTV monitoring, anti-tailgating, multi-factor authentication for entry and many more. Following strict security principle across application, compute, network, perimeter and identity management can definitely make a secure system. But what if any of the layer is breached, then zero trust architecture comes for rescue which is built on concept of no default trust and achieves security by proper authentication at each layer.

Figure 1. Layered security

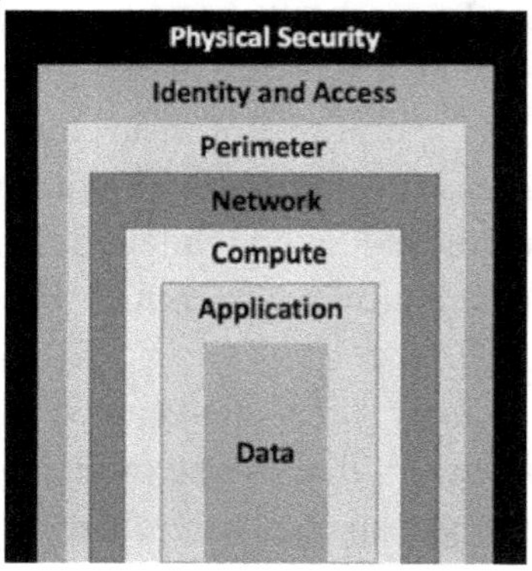

Today's OTT platforms provide best in class Video On Demand(VOD) and Live Streaming services to their customers and earn revenue by providing premium content to users via subscription or by displaying pre-roll, mid-roll or post-roll ads while playing contents to various users who can be guests or unregistered users, registered users or subscribed users. Users can access content anytime, anywhere from various channels like mobiles, web, TVs, set-top boxes, gaming consoles etc over open internet. This in turn creates security challenges in terms of protecting content, protecting user data and privacy and application availability. **Figure 2** depicts a sample OTT architecture which need to be secured following zero-trust architecture principles. As described in the architecture the frontend contains various user devices and Content Delivery Network (CDN) services which are hosted in open network. Content providers provide their contents to OTT platforms which is then stored in CDN for faster access after required transcoding and media format conversions. Content request is initiated from various front-end channels and are served to authorized users with the help of middleware platform. The middleware platform provides various APIs to access catalog, content URL, their metadata, subscription status etc along with third party integrations like payment gateway, sports widgets. The middleware provides complete abstraction to backend services and on successful authentication and authorization contents are streamed to the end user devices. There are various logging and monitoring tools available to capture logs generated from various systems for subsequent analysis and reporting. Most of the OTT providers leverage cloud PAAS and SAAS services to host their applications except few systems which can not be hosted on cloud due to legacy or compliance or unsupported nature. In the following section we have described how security is achieved as a whole for the sample OTT architecture which can also be extended to other domains.

Figure 2. OTT system architecture sample

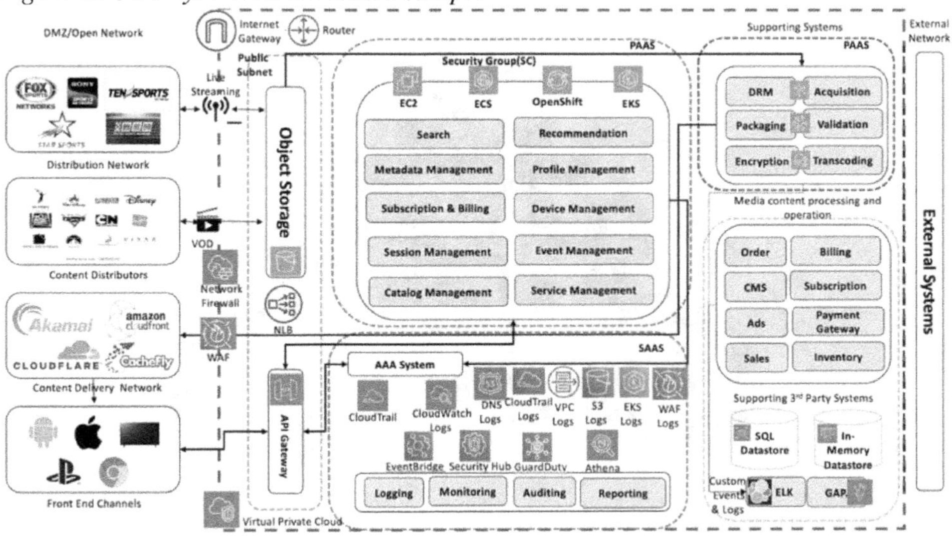

04. SECURITY ARCHITECTURE AND DESIGN REVIEW

Any unauthorized access to OTT content not only impact revenue but also impacts the reputation of the organization. Various security threats faced by OTT systems includes:

- Theft of user privacy information like mobile number, email etc.
- Unauthorized access of content not specific to particular subscription
- Attack on APIs to access authentication tokens for logon and use of the system
- Malicious attack on system infrastructure leading to system down or interrupted service

Existing zero-trust security frameworks usually use subject and object as core security dimensions (Samaniego & Deters, 2018; Zimmer, 2018), but in this chapter we have considered multiple dimensions like infrastructure, application and code, storage and data, content, authentication and authorization, monitoring and control. This framework as given in **Figure 3** is covering all aspects of OTT framework and provides end to end security following zero trust architecture principle.

Figure 3. OTT security framework

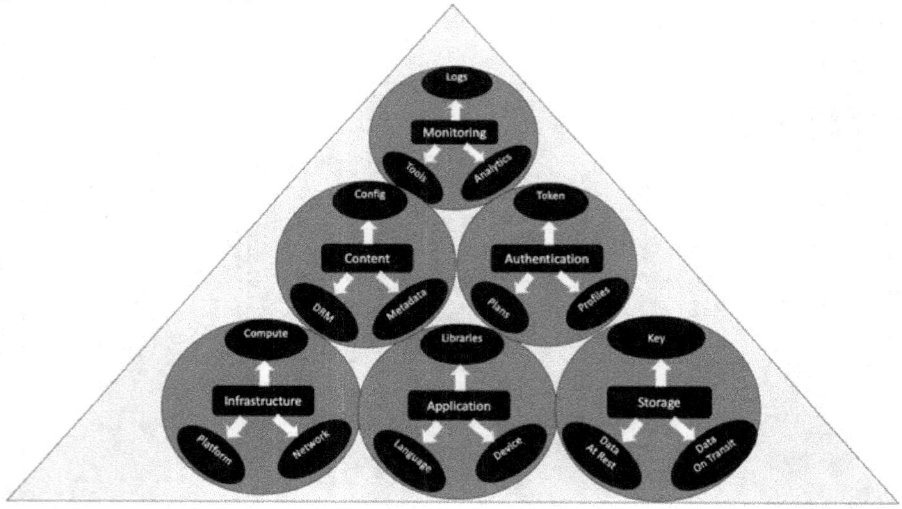

Security Architecture and Review (SAR) as given in **Figure 4** is a process of evaluating system, application, network to identify potential security threats considering design, implementation and deployment. Typical steps for security architecture review contains the following:

- Review of proposed architecture and design
- Threat modeling to identify security threats and vulnerabilities
- Evaluation of security controls and countermeasures as included in the design
- Identify potential flaws in the system
- Review and submit documentation along with recommendations

Figure 4. Security architecture review and design process

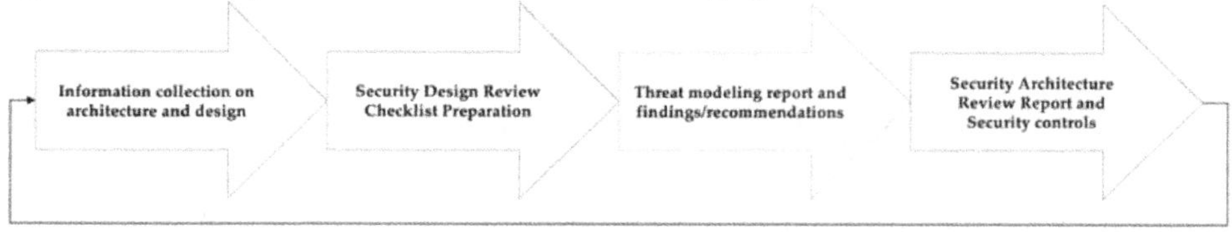

Various threat-modeling tools like Microsoft Threat Modeling Tool, IriusRisk, ThreatSpect, ThreatModeler etc are used to do the threat modeling using several techniques like STRIDE, PASTA and DREAD. While STRIDE categorizes threats on type of attack, DREAD prioratizes security threats based on level of risk whereas PASTA considers the total development cycle for modeling threats. These tools help to identify possible threats for a proposed architecture like spoofing, tampering, repudiation, information leak etc. Microsoft Threat Modeling Tool is used to do the threat modeling of given sample OTT architecture as given in **Figure 5.** which suggests various threats to be addressed.

Figure 5. Threat modeling for sample OTT architecture

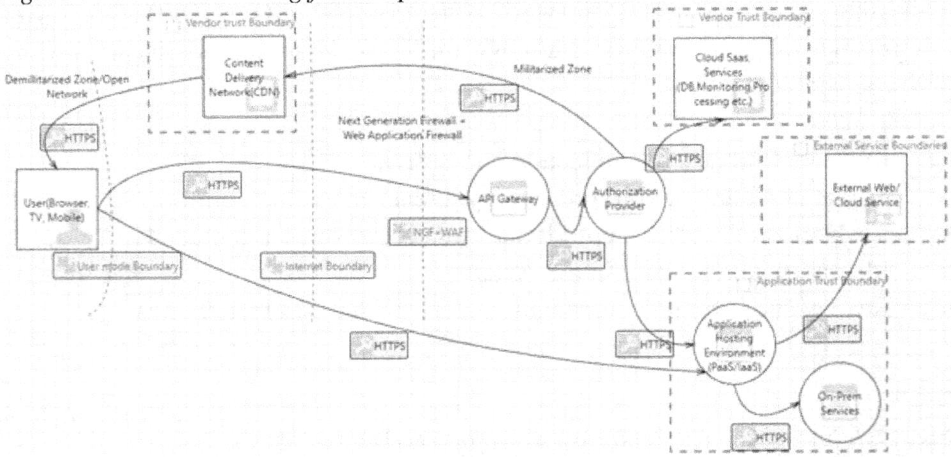

SAR Threat model suggests possible threats that exist in the proposed architecture but don't clearly state how to achieve the same. There are some licensed products who provide the remediation measures in exchange of high licensing cost which is not accessible to all. Following sections give brief overview on various measures to be taken to achieve the zero trust architecture for OTT platform using the OTT security framework as depicted in **Figure 3.**

05. INFRASTRUCTURE SECURITY

All most all OTT platforms leverage public cloud infrastructure or hybrid cloud infrastructure to reduce capital expenditure (CAPEX). 'Pay As You Go' model helps to scale their infrastructure with the increase in subscribers and in turn manage their operation expenditure. Irrespective of the platform they choose, following are the key considerations while managing cloud infrastructure with respect to security.

- Managing identity
- Managing authentication and authorization
- Managing application running on cloud infrastructure
- Implementing security for virtual network
- Implementing security for public access to resources

While the implementation may differ across various cloud providers, conceptually all of them follow standard security best practices to keep their infrastructure secure. It is advisable to disable any security defaults and configure as per need. Role Based

Access Control (RBAC) must be used to manage resources based on assigned roles. Identity provider like Azure Active Directory in Azure and Identity and Access Management (IAM) in AWS helps to manage both predefined and custom roles. Authentication and authorization can also be managed by these identity providers. Following best practices are to be followed to achieve tighter security goals.

- Follow the principle of least privilege.
- Disable default security provided by providers and enable security as per need.
- Users, Groups and Roles to be managed following RBAC policy.
- Conditional Access and Multi Factor Authentication (MFA) features like authenticator app or phone based service to be enabled for secure authentication. Ensure device trust for limited period in case authentication is enabled for trusted device.
- Ensure to validate user by possible maker checker process during creation.
- Ensure to define proper password creation, rotation and expiration policy.
- Its recommended to encrypt data both at rest and transit.
- Resources are to be kept under specific network boundaries or subnets with clear defined ports for communication.
- Use the key management service or vault service to store sensitive data like username, password, keys etc.
- Application access to resources is to be done via access token and bearer token as available in standards like OAuth2.0.
- Ensure to keep sign-in logs, audit logs, provisioning logs etc for future analysis.
- Identity protection if available to be enabled which can remediate user-risk of leaked credential or sign-in risk which includes anonymous ip access, trying out different password, device infected with malware etc along with required controls.

Computing resources such as Virtual Machine (VM) are completely managed by cloud platform. However, placing this as part of virtual network is the responsibility of administrator. Network security groups or rules should be carefully configured to allow traffic to and from specific port for the VM. Most of the cloud providers provide various reusable custom groups which can be applied to any VM, but configuration of the same lies with the administrator. It is also to be checked the images used to create VMs must be trustworthy. It is recommended to create separate virtual networks for different user access like public access, private access within network or best on type of application like front-end, middleware, back-end etc. Communication between networks should be through properly defined ports by applying required network policies. RBAC principles should be followed while granting access as given in **Figure 6.**

Figure 6. Role assignment in Azure

For networks, Network Security Group (NSG) in azure and Security Group in AWS plays major role of restricting access to network. Logically, networks are further sub-divided into subnets to host group of applications like frontend, middleware, DB etc and appropriate permission on IP or HOST or Port are to be given. Appropriate malware protection software can be installed for the virtual machine or other resources deployed in this virtual networks. Regular security updates or patches need to be installed in VMs. It is recommended to use latest image while creating new VM. Options like Microsoft Defender for Cloud or Amazon GuardDuty can be used to detect and monitor active threats in VMs. Where applicable Just in Time VM access to be given to VM. Disk encryption to be enabled for encrypting the disks. While security group rules restrict inbound and outbound access, it is recommended to have continuous review of the same. In some cloud providers like Azure application security groups provide additional security restriction on application access. So security group for specific platform need to be analyzed carefully to implement tighter security policies. While connecting multiple networks or data center to cloud VPN solutions can be explored and based on the requirement a Site to Site VPN or Point to Site VPN connection can be established. This ensure the network traffic across data center remains secure.

There is always possibility of some software components running in data center and it is connected to applications running in cloud either due to deployment constraint, regulatory purpose or organization policies. Servers running in data center need to be hardened to reduce attack surface, few of which is as given below;

- Server communication should be encrypted
- Removing telnet, ftp, remote login services
- Removing unnecessary packages or libraries
- Updating operating system and software regularly to latest one
- Enabling password lockout policy and disabling root access
- Disabling unwanted services
- Separating operation system files via disk partitioning
- Disabling USB and other external ports
- Configuring system firewalls
- Enabling logging, monitoring and auditing

06. MIDDLEWARE APIS AND CONTAINER SECURITY

Enterprises now have moved from legacy bare metal based hardware to cloud based 'Pay As You Go' virtual servers with unlimited processing, memory and storage capability which can scale on demand. This has led to microservices based architecture where services are deployed in containers as an isolated server. Orchestration platform like Kubernetes provide the required infrastructure to host cloud native applications following a microservice based architecture and security of Kubernetes cluster is one of the critical factor in securing APIs.

Cloud platforms normally expose metadata services locally to nodes and by getting access to pods sensitive data can be accessed by attackers. To secure node metadata following best practices are notable:

- Disabling legacy metadata APIs
- Creating service accounts with minimum permission
- Accessing VM metadata using non routable IP address
- Leveraging node selector, label and taint options while pod scheduling
- Encrypting secrets at rest
- Using network policies to restrict pod access
- Restricting alpha or beta feature access
- Continuous monitoring

Endpoints in Kubernetes expose services and to secure endpoints following best practices are generally followed:

- Putting firewall in front of load balancer
- Enabling Transport Layer Security for all traffics for API
- API authentication and authorization

- Disabling public endpoint access
- Restricting access to *etc* database
- Reviewing third party integrations
- Controlling access to kubelet

Access to Web UI dashboard is also to be disabled and in case access required it can be done via token. But it is recommended to disable Web UI dashboard in production and use kubectl commands to interact with cluster. Its strongly recommended to use Center for Internet Security(CIS) benchmark to secure master and worker nodes which states what are different permission to be given to various configuration files and who will be the owner of it. Control plane in Kubernets cluster exposes the inherent APIs for which following best practices may be followed:

- Enabling TLS for API server
- Disabling anonymous requests to the API server
- Discarding basic authentication
- Rate limiting API server requests
- Retaining and rotating logs
- Refraining binding to insecure ports
- Using certificate for authentication

Along with these to make the cluster secure, other practices like frequent update of kubernetes, hardening kernel tools, minimizing external access to network, minimizing Identity and Access Management (IAM) roles, minimizing base image footprints, enabling Mutual Transport Layer Security (mTLS) for pod to pod communication, scanning images for known vulnerabilities along with monitoring audit logs are few proven principles to achieve higher level of security. While using container images it is advised to sign the image and upload in container registry to avoid any malware injected by others. While above best practices make the orchestration platform secure, microservices running on this platform need to be secured. Following are the best practices to achieve microservices security:

- Protecting entry point to the services
- Securing service to service communication using certificates
- Each service should be deployed in its own container with own entry point
- API gateway between external world and services
- Usage of bastion host between system and services
- Following Roal Based Access Control (RBAC)
- Applying network policy across group of services
- Using namespace to restrict service access

- Implementing mTLS
- Building trust boundaries
- Following a token based policy for service access
- Using service mesh

Containers host the microservices and create attack surface, following best practices can be used to secure containers:

- Managing OS permissions
- Namespace and Cgroups in namespace
- Hardening the container host machine
- Role Based Access Control
- Signing container images
- Scanning container images
- Isolating containers
- Using encrypted communications
- Protecting container runtime
- Disabling root access and using secure images
- Checking binaries using checksum before deployment

07. APPLICATION AND CODE SECURITY

Application security is one of the major pillar of zero-trust architecture framework and need to be taken care even if not stated explicitly. During the Software Development Lifecycle (SDLC), security requirements need to be captured during requirement gathering phase and during design phase both the application architecture and threat modeling via SAR is carried out. During coding its necessary to follow secure coding methodologies, doing security testing. During testing phase penetration test is carried out to find out the gap. Code once developed need also to be deployed securely with safe configuration management principles. Various mitigation features like authentication, integrity, non-repudiation, confidentiality, availability, authorization etc are followed to mitigate various threats. Several tools like Static Application Security Testing(SAST) analyzes the static source code and Dynamic Application Security Testing(DAST) helps to analyze run time applications for existing vulnerabilities. Open Web Application Security Project (OWASP) in an open community which help organizations to build, run and purchase secure and reliable applications. OWASP provides top 10 security risks which must be handled during application development along with various guides for handling security issues. This section focuses on high level guidelines towards making the code secure.

Vulnerability Assessment and Penetration Testing (VAPT) is done on developed software or application stack to identify various threats. OWASP web security testing guide with any standard tool like HTTPTracker, Fiddler, Burp, Postman etc can be used to carry out VAPT test on the application and identified risks are mitigated. There are also vulnerability scanners like Nikto, Nessus who have the capability to scan the web server and find out possible threats. In a nutshell VAPT activity is must for any application to make it secure as it identifies various attack surface for web applications like

- Session hijacking
- Improper modification of parameters
- Denial of Service
- Cross Site Request Forgery
- Cookie modification
- SQL Injection attacks
- Cross Site Scripting attacks
- File Injection attacks
- Command Injection attacks
- Exposing source code
- Directory traversal attack etc.

OWASP coding guideline is great source for addressing above issues. There are cheat sheets for various attacks like sql injection, authentication, session management etc. which suggest how to address security issues during coding. Following are some of the best practices followed for application security

- Doing code reviews
- Static code analysis
- Securing configuration management
- Application white listing
- Ensuring transport layer security (TLS/SSL)
- Strong authentication and authroization
- Encrypting application configuration data
- Granting appropriate application access
- Removing default accounts
- Scanning code and images for vulnerability
- Tuning database query

Mobile devices based on android and iOS platform provide the required interfaces to play OTT content and these applications need to be secured too. These applications are mostly native applications which are built on top of android or xcode platform or can be hybrid applications built on some standard frameworks like react native. Irrespective of platform they are built in, standard coding practices need to be followed for each. These platforms stores or caches key login and user data within to improve application performance. Clear strategy to be devised when the local cached data will be expired and any access to API for subsequent authentication and catalog access will be done following the standard token based authentication as given in authentication and authorization section. The code should be secured with appropriate certificate and following best practices as per Google Play store and Apple App store. Google and Apple both scan the application builds when deployed in their store and notify developers about possible libraries and vulnerability. Apps are also traced and checked via third party application like Firebase. It is recommended to use standard secure SDKs while building these applications. It is also to be ensured that no Personal Identifiable Information (PII) data is stored or exposed without user consent. It is recommended to encrypt PII data before storing. Following are the best practices for mobile applications running the OTT contents:

- Encrypt sensitive data while storing on client side
- Implement API security via API keys or token following standards like OAuth
- Secure HTTP channel with help of TLS for achieving confidentiality, integrity and authentication.
- Preventing API call tampering by separating API key into ID and a shared secret key is passed in HTTP header and secret key is used to sign or encrypt information in transit.
- Handling user credentials and keys with the help of keychain service for iOS and Shared Preferences for android. Authentication should be a combination of client's private TLS key and API secret.
- Suggest to follow some Mobile Security Framework available for best practices.

08. STORAGE AND DATA SECURITY

Any enterprise including OTT platforms need to manage the secrets. Secrets can be encryption keys to encrypt the data, certificates to communicate with other systems, password to connect a database or a key to connect to storage and many more. To ensure highest level of security it is of great importance to manage and store the secrets effectively. Key Vault services provided by various vendors or cloud providers help to achieve the objective. Azure Key Vault Service and AWS Secrets Manager are two of the widely used managed services provided by Microsoft and Amazon to manage the secrets. While accessing third party service outside network service endpoints can be created to manage connection.

Data in any platform is at the end transferred via a network and stored in disk. While data is moving in a network like a fetch request from database there is the probability of man in the middle attack. Similarly, when data is stored in disk which physically lies in a data center need to be preserved from unauthorized access. Although, data centers are secure and various security measures taken in terms of accessing the disk, still additional measures required to secure the data at a time if the disk is accessible to unauthorized user. Encryption in transit and encryption at rest are the two concepts which ensures data is secure during transit in network and while stored in disk with the help of encryption keys.

Most of the cloud providers provide option to enable server side disk encryption which ensures data at disk is encrypted. There are options to use either platform managed key or customer managed key to do the encryption. All these keys are managed in the key vault service as shown in **Figure 7**. It is advisable to rotate keys available in azure key vault. It is recommended to use SSL certificate for internal and external communications too.

Figure 7. Key Valut in Azure and AWS

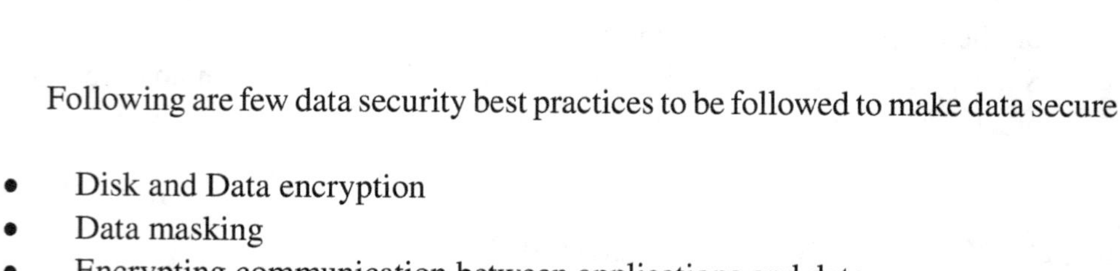

Following are few data security best practices to be followed to make data secure

- Disk and Data encryption
- Data masking
- Encrypting communication between applications and data
- Regular backup
- Data auditing
- Real time alert
- Purging stale data

- Adhering to standards and government regulations

09. CONTENT SECURITY

Content in OTT mainly refers to video files which need to be played back on demand or on live stream. Contents are one of the major source of revenue both for OTT providers and content creators and need to be secured from unauthorized users or subscribers. Digital Rights Management (DRM) systems play major role in securing content and restrict content duplication, distribution and viewing. Contents are encrypted by these systems and can only be decrypted via appropriate key which is only exchanged to authorized user and devices. High level steps of DRM as shown in **Figure 8** is described as below

1. Content is encrypted using specific key and stored in Content Distribution Network (CDN).
2. User logs on to OTT system via proper authentication and get access to content catalogue.
3. User then request for video playback from a specific device and both content id and device id is passed to the system.
4. Contents get downloaded from CDN system based on the playback URL defined.
5. Request is then made to DRM proxy server by passing content id, device id and content key which is then passed to internal DB. If authorization is successful, key is returned as a response to player.
6. Returned key is used to decrypt the video and playback happens with help of player.

Figure 8. Sample DRM architecture

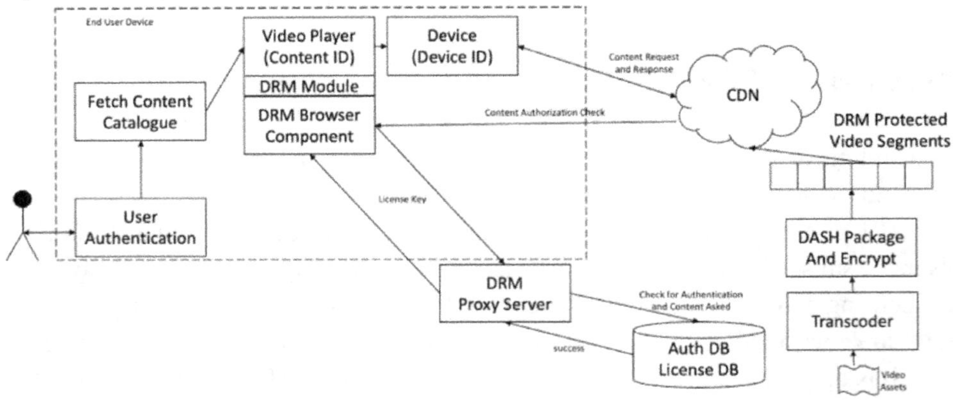

Video content played by OTT providers can be Video On Demand (VOD) which is secured by DRM systems or direct live feeds. Http Live Streaming (HLS) is one of the popular technology developed by Apple for live streaming content. There are other protocols too like Dynamic Adaptive Streaming over HTTP (DASH) and Real-Time Messaging Protocol (RTMP). Streaming happens by transferring chunk of videos of specific bitrate based on the bandwidth and quality chosen with the help of playlist or manifest file. These files are maintained in Content Management System (CMS) along with other metadata and secured via API authentication.

10. AUTHENTICATION AND AUTHORIZATION

Authentication and authorization is the first line of defense for getting hold of data or various resources running in the eco-system. Services deployed in cloud already manages the same by using various identity management solution as described in detail in infrastructure section. But in real world, enterprises may follow a hybrid approach where some applications or services run in their data center and they too access or interact with third party applications running outside of their secure network. Standards like OAuth and OpenID Connect (OIDC) comes here for rescue where the authentication to third party happens via token exchange in stead of following legacy approach of password sharing. This also is a standard way of securing APIs, mobile apps or single page applications accessing data from outside world or open network.

Simple applications accept username and password to achieve basic authentication, but things become messy when authentication is done via mobile, smart tv or while implementing single sign on where standard username and password is used across applications. Password hashing a technique suggested above also do not guarantee security as passwords are exchanged with external applications. Even implementing MFA across applications do not scale well and hence OAuth is the standard way of implementing authentication across various applications, APIs, devices etc in a more flexible and scalable way. The authorization server checks the identity and hands over a token which is used to authenticate the application or requester for a particular resource called resource server. OIDC works on top of OAuth by adding user identity using the same secure principle. In simpler terms, if OAuth is used for accessing APIs then OpenID is used to identify user using the secure principle of exchanging token with validity.

In OTT platforms as shown in architecture, almost all data is exposed to various channels via APIs and API security is of paramount importance. Following the terminology of OAuth the user becomes resource owner, the device act as user agent, the application acts as client, the token is generated by authorization server to access the API which acts as a resource server. Authentication only happens with the authorization server where the application has no information on user authentication parameters. Tokens can be reference token which is just a random string and refers to some data in a database or cache store with additional information or it can be self contained token where the token has all required information. Each having their pros and cons but nevertheless following the token management principle of OAuth and any standard authorization server supporting OAuth and OIDC the middleware application can be secured and authenticated for each request. While reference tokens are easy and simple to build and stored along with other required information it also gives the access to developers to introspect the token, also storage space may grow keeping in view of number of users, validity of token, additional information kept along with token. This may sometimes put overhead of network lookup if the authorization server is not built in or not sharing database. The most commonly followed approach is to use self contained token using Json Web Token (JWT) which contains all information with no requirement of shared storage and network access. Often JWT token is not encrypted and may put some risk of exposing user data so need to be handled carefully. It also remains valid till the expiry period even if the user or application is deleted. Its recommended to use some sort of middleware or API gateway which can do local validation of tokens and can easily eliminate malformed, expired tokens etc boosting performance. The API can then act as second layer where they can do token introspection of revoked and valid tokens and based on that provide access. In a nutshell, its advisable to use Identity provider solutions provided by cloud platforms for managing cloud resources along with OAuth based token management principle to validate internal or third party requests hosted outside cloud to achieve the best of security.

11. MONITORING AND CONTROL

Monitoring and control is one of the critical activity in zero trust architecture. It helps to perform real time analysis and do required audits to ensure everything planned is in place. Tools like Azure Monitor and AWS CloudWatch as given in **Figure 9** have the ability to collect various metrices of the resources deployed and generate alert as per the rule defined. It also helps to analyze various activity logs and get application insights by collecting logs and storing them in central location. In some enterprises the collected logs are ingested to separate Security Information and Event Management (SIEM) system so that it can be analyzed for anomalies. Tools like EleasticSearch, FluentD and Kibana (EFK) provide features for log aggregation and visualization options for easy monitoring. These tools have the flexibility to ingest logs to central ElasticSearch repository and do the visualization via custom graphs using Kibana.

Figure 9. Sample Monitoring dashboard in Azure and AWS

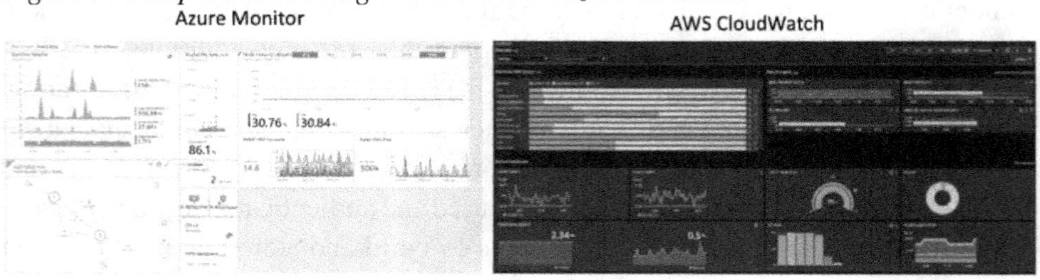

Container platforms like Docker, Kubernetes etc. also provide various plugins to monitor the state of services. Service mesh platform like Istio which can run on container orchestration platform provide multiple features for observability, logging, security, traffic management etc. Kiali dashboard available as part of service mesh can show the traffic routing, service status along the mTLS connection for security. Open tracing feature of Jaeger UI provides the detail tracing along with execution time. This helps which service is calling which other service methods and helps to monitor service endpoint and their response as shown in **Figure 10.**

Figure 10. Sample Kiali dashboard and Jaeger UI

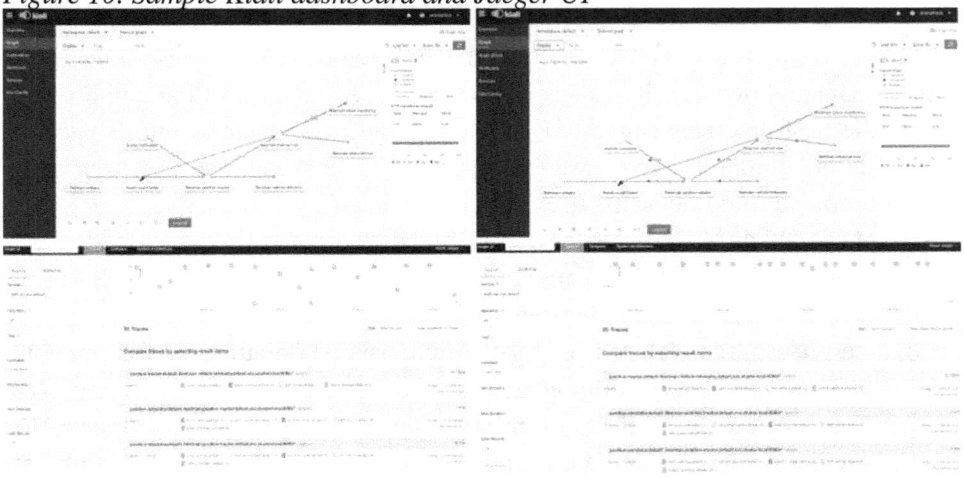

EFK is also available as part of service mesh extension as shown in **Figure 11.** In some enterprises application logs or other system logs are ingested to a custom EFK solution and custom dashboards created for monitoring and control. Irrespective of the toolset, OTT platforms must ensure to collect logs from various sources like VMs, Applications, DBs, Logon Activities etc and create appropriate rules to send alert to security team in case of any deviation. Machine learning models can also be integrated to ensure automatic threat or anomaly detection based on log analysis.

Figure 11. Sample Monitoring dashboard of Istio service mesh

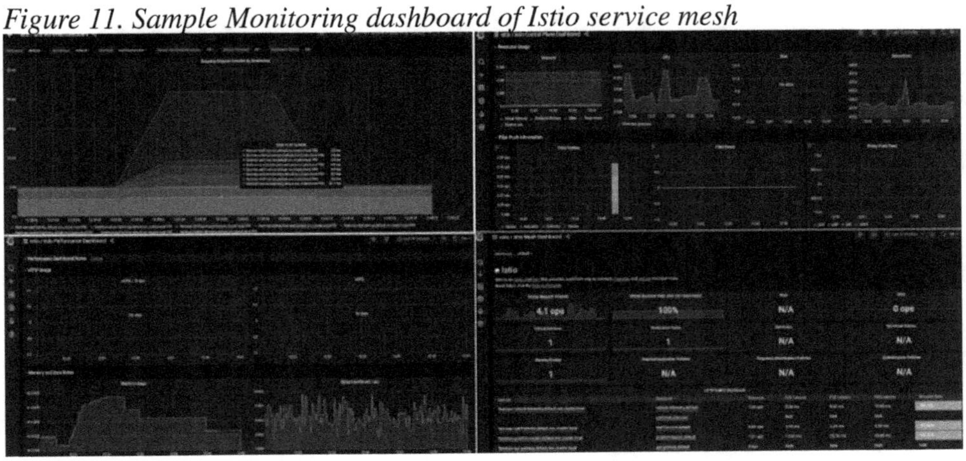

12. CONCLUSION

In this chapter we briefly describe the proposed OTT security architecture framework and how security issues can be addressed on each of the components. While the best practices suggested on each of the component is not exhaustive, but definitely comprehensive enough to achieve security on that particular layer. We recommend to do Security Architecture Review (SAR) during architecture or design phase and do VAPT after the application development. We discussed briefly on the usage of SAST and DAST tool in identifying bottlenecks along with various available authentication and authorization mechanism.

In a zero trust architecture principle, there exists no default trust for any component and each and every component in the eco-system need to have some clear distinct way to validate itself to get the required authentication and authorization on a particular resource. Communication between the components need to be secured too. Following zero trust architecture principle can prevent the attacker to gain unauthorized access to any component or data or resource and minimize the impact on the enterprise. OTT being accessed from various channels like mobile, TV, set-top boxes etc on open internet, its critical to secure the data and content to avoid reputation and revenue loss. Although, we have not covered much on the security of third party applications which are outside the enterprise scope, we have discussed in detail how token management using OAuth comes to rescue for secure authentication and authorization. We apprehend that by following the suggested framework principles and best practices, any OTT provider can achieve the highest level of security for their platform.

REFERENCES

Chen, B., Qiao, S., Zhao, J., Liu, D., Shi, X., Lyu, M., Chen, H., Lu, H., & Zhai, Y. (2021). A security awareness and protection system for 5G smart healthcare based on zero-trust architecture. *IEEE Internet of Things Journal*, *8*(13), 10466–10476. DOI: 10.1109/JIOT.2020.3041042 PMID: 35783535

D'Silva, D., & Ambawade, D. D. (2021). Building a zero trust architecture using Kubernetes. In *Proceedings of the 6th International Conference for Convergence in Technology (I2CT)*. DOI: 10.1109/I2CT51068.2021.9418203

Goodman, A. (2019, June 25). *What is Zero Trust?* DZone. Retrieved from https://dzone.com/articles/what-is-zero-trust

Open Web Application Security Project (OWASP). (n.d.). *OWASP*. Retrieved from https://owasp.org/

Pandey, P., Mishra, S., Rai, P., & Anand, A. (2019). Social engineering and exploit development. [IJSRCSAMS]. *International Journal of Scientific Research in Computer Science Applications and Management Studies*, *8*(5).

Rose, S., Borchert, O., Mitchell, S., & Connelly, S. (2020). *Zero trust architecture* (NIST Special Publication 800-207). National Institute of Standards and Technology. DOI: 10.6028/NIST.SP.800-207

Samaniego, M., & Deters, R. (2018). Zero-trust hierarchical management in IoT. In *Proceedings of the IEEE International Congress on Internet of Things (ICIOT)* (pp. 88–95).

Statista. (n.d.). *Revenue from OTT and video across India from 2018 to 2022, with estimates until 2027*. Retrieved from https://www.statista.com/statistics/1231740/ott-and-video-revenue-india/

Wu, Q., Wu, C., Yan, X., & Cheng, Q. (2021). Intrinsic security and self-adaptive cooperative protection enabling cloud-native network slicing. *IEEE Transactions on Network and Service Management*, *18*(2), 1834–1848.

Zimmer, B. (2018, January). LISA: A practical zero trust architecture. In *Proceedings of Enigma*, Santa Clara, CA, USA.

Chapter 5
Cryptocurrency Price Prediction:
A Comprehensive Analysis of Social Media Sentiment Analysis

Nrusingha Tripathy
https://orcid.org/0000-0002-0272-7479
Siksha 'O' Anusandhan, India

Debahuti Mishra
Siksha 'O' Anusandhan, India

Sarbeswara Hota
Siksha 'O' Anusandhan, India

Subrat Kumar Nayak
https://orcid.org/0000-0002-7438-9085
Siksha 'O' Anusandhan, India

ABSTRACT

The world of cryptocurrencies has expanded rapidly over the last ten years, with the most recent advancements being witnessed. According to Twitter statistics, there are roughly 1,500 tweets on Bitcoin alone per hour, which lends credence to this claim. In this work the weekly dataset for Bitcoin (BTC) and average weekly sentiment data for Bitcoin (BTC) are considered. This work employs four distinct models: Naive Bayes, valence aware dictionary and Sentiment Reasoner (VADER), Gated Recurrent Unit (GRU), and Long Short-Term Memory (LSTM). The LSTM model performs well and has a greater accuracy of 0.952 than other models. Text sentiment is frequently impacted by intricate, non-linear word-phrase interactions.

DOI: 10.4018/979-8-3693-8643-9.ch005

These complex patterns are captured by LSTMs, resulting in more precise sentiment analysis. By monitoring real-time tweets, we can predict the sentiment of the Bitcoin market. The LSTM model gains the ability to forecast price changes based on the mood conveyed in tweets by training it on past tweet data and matching fluctuations in the price of Bitcoin.

1. INTRODUCTION

There is a growing need for a more transparent and easily available kind of cash in the overly-marketed world of today. The idea of digital money has expanded since the introduction of the first virtual currency, Bitcoin, in 2009 as the world's technological capabilities and the realm of financial transactions continue to progress. The idea behind cryptocurrencies is to facilitate safer and quicker online financial transactions (Jung, Lee, Lee, & Kim, 2023). This is made possible by a technology known as the Blockchain, which makes transactions significantly safer since there is no single entity to validate them. Instead, the system is decentralized. This turns out to be far more beneficial than the existing system for verifying and completing fiat money transactions, where some of the worst financial frauds of the century have resulted from the ease with which they may be abused (Ramani, Jahnavi, Reddy, VenkataChakravarthi, Meghanath, & Imran, 2023).

Using encryption to protect unit production and transactions, cryptocurrencies have grown in popularity quickly since its launch in 2009 with the establishment of Bitcoin. A lot of people use cryptocurrencies as a means of engaging in high-risk, high-reward short-term investment rather than as a sincere attempt to overthrow established financial institutions due to their extremely unpredictable market. By doing sentiment research on the social media site Twitter, this study intends to explore one of the possible causes of this volatility: the general sentiment toward Bitcoin (Saleem, Yaqub, & Zaman, 2024). The goal is to see if there is a link between the market value of Bitcoin and the mood towards it. The initiative also intends to look into how tweet popularity could affect how strong this association (Doroslovački, Gradojevic, & Tarnaud, 2024).

The conclusions drawn from this research may help investors both new and old make well-informed decisions that follow market trends and increase their returns on investment. To find any noteworthy patterns or commonalities among trends in the two variables, the strength of this association was assessed using manual data analysis and time-series visualisation (Arslan, 2024). According to the findings, there could be a relationship between the mood surrounding Bitcoin and its value swings, which might be used to forecast the direction of the market. However, this relationship does not account for the size of fluctuations in the Bitcoin market.

For Bitcoin sentiment analysis, this work uses four different models: Naive Bayes, VADER, GRU, and LSTM. Since rapid changes in sentiment have a big influence on pricing in the cryptocurrency market, LSTM models are well-suited to monitor and process massive amounts of sentiment data instantaneously. Sentiment analysis is used by traders and investors to help them make better-informed decisions. An understanding of market mood may be useful in reducing risks and identifying potential opportunities for purchases or sales. Because of the unique benefits and capabilities of each model, a comprehensive analysis of the sentiment from scraped tweets is made feasible. Sentiment analysis is a useful tool for traders to employ when making decisions. A positive mood might suggest that now is a good time to purchase, whilst a negative attitude would suggest that there may be a sell-off.

2. LITERATURE REVIEW

Forecasting Bitcoin sentiment analysis has several advantages that can improve risk management, trading tactics, and general market comprehension. Sentiment research sheds light on the attitudes that traders and investors have toward Bitcoin at any particular moment. This can aid in the prediction of changes in investor behaviour as well as fluctuations in the market. Here are some summaries of earlier work.

According to Jung et al. (2023), because Bitcoin price patterns reflect the general trajectory of the cryptocurrency market, it is vital to predict them. Studies on the variables influencing variations in Bitcoin prices have been carried out because of the market's brief history and high level of price volatility. Using material from Twitter, experiments have been done to forecast the price of Bitcoin. But there wasn't much data available, and price predictions were made for a very short time (less than two years). Data from Reddit and LexisNexis, spanning more than four years, were gathered for this study. By combining technical and psychological indicators with the price data and post volume, these data were used to quantify and evaluate the efficacy of the six machine learning algorithms.

Ramani et al. (2023) describes Bitcoin is the most widely used cryptocurrency worldwide and allows users to conduct safe online transactions. This helps to keep your money private when doing fast transactions, such as cash transactions. In recent years, the majority of consumers have shown interest in the Bitcoin ecosystem. Because of its extreme volatility, precisely predicting the price of Bitcoin is a challenging undertaking. The primary problem with Bitcoin is its price volatility, which highlights the need of comprehending the underlying pricing methodology. The primary variables affecting the price of Bitcoin include availability, supply and demand in the market, the presence of competing cryptocurrencies, and investor sentiment. The goal of this research project is to create a reliable and accurate model for efficient Bitcoin price prediction.

Saleem et al. (2024) describes a novel framework that combines essential components of prospect theory with the core ideas of electronic word-of-mouth (EWOM) to anticipate changes in the price of Bitcoin/USD using sentiment analysis on Twitter. Our study has significant ramifications for both investors and risk managers, providing an essential tool for informed choices in the Bitcoin market. We mix precise positive and negative sentiment values instead of relying solely on a single compound score, which is different from typical techniques. In particular, our novel method for sentiment research separates sentiment into its favourable and unfavourable facets, enhancing our comprehension of the changing patterns of market mood. It also provides investors and financial companies with a deeper comprehension of the risks associated with assets other than Bitcoin that are affected by sentiment-driven market dynamics.

According to Doroslovački et al. (2024), the collective attitude or conviction those who participate in a financial market hold regarding the market or an investment is referred to as “MARKET sentiment.” It is commonly gauged by investors' degree of optimism or pessimism on the market or the security in question's future prospects. A multitude of factors, such as economic statistics, political developments, and market news, can impact market attitude. Given that it may significantly affect both the need for commodities and the price at which they are traded, it is crucial for traders and investors to take this into account. In a time-series setting, the overall emotion of the financial sector may often be summed up by adding up the subjective opinions of both its knowledgeable and ignorant players, or traders. There would be a broad range of beliefs if every trader operated freely and according to their own views, which would lead to a relative seasonality of market pricing.

Arslan (2024) labels due to increased investment, cryptocurrencies have recently attracted a lot of attention. In this study, we investigate the relationship among Twitter activity and Bitcoin prices by looking at tweets on attitudes surrounding the price of the cryptocurrency. We offer a model with two different networks. The empirical mode decomposition approach is used to further break down historical pricing data into several components, which is the only source of data used in the initial network. This breakdown lessens the effect of erratic swings on estimates of the price of Bitcoin. Due to increased investment, cryptocurrencies have attracted a lot of interest lately. In addition, social media has gained popularity among scholars as a prediction tool, especially when it comes to financial markets. Twitter and other platforms are used by investors for trend analysis and investment analysis, which can have an immediate effect on Bitcoin's future price swings. Comprehending and examining the mood on Twitter may offer valuable perspectives on forthcoming fluctuations in the price of Bitcoin and illuminate the ways in which investor sentiment impacts cryptocurrency markets.

In this work four different models are used for Bitcoin sentiment analysis. Early indicators of trend reversals can be found with the use of sentiment analysis. Sentiment fluctuations, whether positive or negative, can occur ahead of real price movements, giving traders a useful early warning system. Sentiment analysis facilitates the tracking of possible volatility and market sentiment. This can help in modifying risk management tactics, such position size and stop-loss orders, to reduce risks during highly uncertain times.

3. RESEARCH METHODOLOGY

The fact that Bitcoin price movements reflect the general direction of the cryptocurrency market, must be predicted. Many research has been done on the variables influencing fluctuations in Bitcoin prices because of the market's short history and high level of price volatility (Tripathy, Hota, & Mishra, 2023). The entire workflow diagram is given in Figure 1. To get current insights on market sentiment, real-time sentiment analysis may be performed using deep learning models. In the quickly evolving Bitcoin market, where opinions may change drastically, this real-time functionality is critical. Deep learning architectures vary widely, which makes it possible to test and choose the model that best fits the particulars of the sentiment data and the objectives of the study (Dong, Li, Jiang, Wu, & McClure, 2020).

Figure 1. Workflow diagram

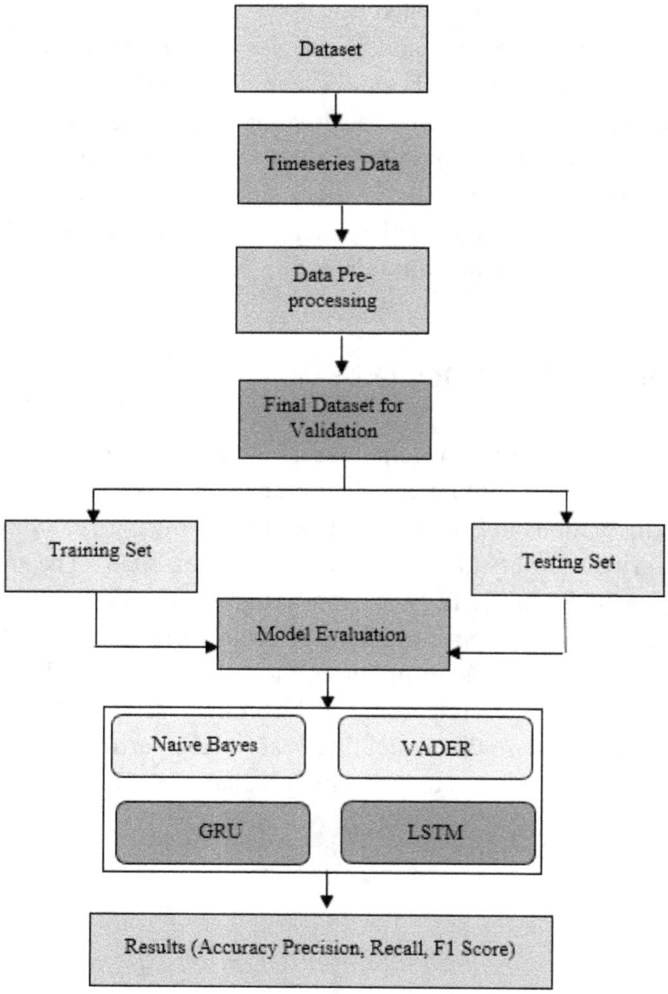

I. Data collection

This study presents a system that makes use of both technical indicators and sentiment indexes. In this work the weekly dataset for Bitcoin (BTC) and average weekly sentiment data for Bitcoin (BTC) are considered. Reddit was the source of social media data, while LexisNexis provided news data. By implementing a sentiment index with weekly posting volume to the available historical price data and several technical indicators, investor sentiment was also taken into consideration. To find pertinent conversations and emotions, concentrate on subreddits about cryptocurrencies. Utilize sentiment analysis methods (such as VADER, LSTM, and GRU) on Reddit data to measure the sentiment conveyed by posts and comments. Sort the sentiment into three categories: good, negative, and neutral. Construct a sentiment index by weekly aggregating sentiment ratings from news sources and Reddit. To take into consideration the amount of activity and engagement, add sentiment ratings to the posting volume.

II. Methodology

Crucial phases in Bitcoin sentiment analysis include statistical analysis and data preparation. The market value of cryptocurrencies has recently surpassed $1 trillion. In this sense, rules about cryptocurrencies, such as Bitcoin, are progressively being developed and they are being acknowledged as a class of assets. Prior research indicates that the use of cryptocurrencies can be applied to portfolio hedging and diversification (Feng et al., 2019). To manage variants of the same term, we first handle text normalization by decreasing words to their base or root form. Effective encoding management in Bitcoin sentiment analysis guarantees reliable processing of text data with non-ASCII characters and multilingual content. Figure 2 shows the average weekly sentiment. We assess sentiment from a variety of textual materials, such as social media platforms like Twitter, by utilizing Unicode encoding, suitable filtering techniques, and model compatibility, successfully capturing the subtleties of emotion conveyed by users (Liu, Cai, Wang, & Lu, 2019).

Figure 2. Average weekly sentiment

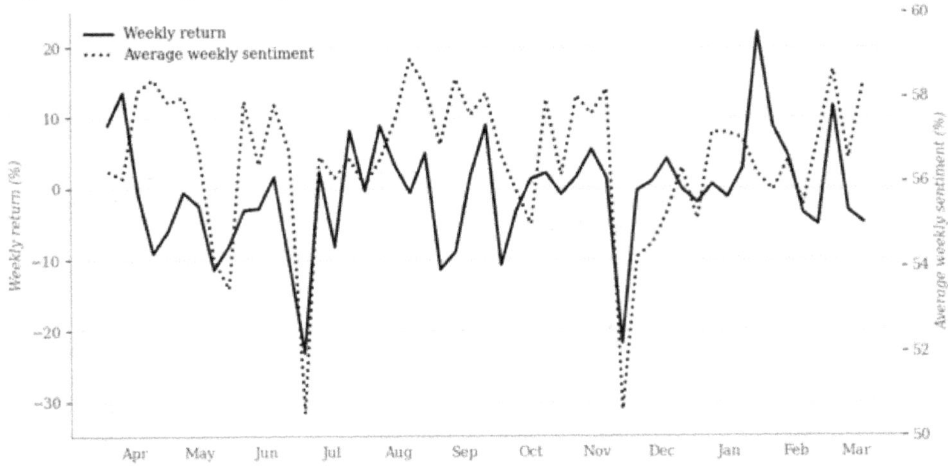

III. Forecasting Method

A. Naive Bayes

Naive Bayes is a computationally efficient and straightforward method to use. It can deliver baseline sentiment analysis findings rapidly and performs effectively with massive datasets. The basic structure of our Naive Bayes model is shown in Figure 3, where A is the class node and also the parent of all nodes. Equation (1) and (2) are the basic Naive Bayes formulation. Each attribute Q_t has a hidden parent Q_{uvt},

$$P(Q_1,\ldots,Q_n,C) = P(C)\prod_{i=1}^{n} P(Q_t|Q_{uvt},C) \quad (1)$$

$$P(Q_t|Q_{uvt},C) = \sum_{t=1}^{n} W_{it} * P(Q_i|Q_t,C) \quad (2)$$

Working with huge datasets of tweets or news items on Bitcoin is advantageous since the model can be trained rapidly and with fewer processing resources. Because of their memory and temporal efficiency, Naive Bayes classifiers are appropriate for real-time sentiment analysis, when speed is of the essence (Tripathy, Hota, Prusty, & Nayak, 2023). Effective handling of large-scale data is necessary for assessing the massive amount of textual data linked to Bitcoin that is produced every day. Figure 3 shows the basic edifice of the Naive Bayes model.

Figure 3. The Naive Bayes model's basic structure

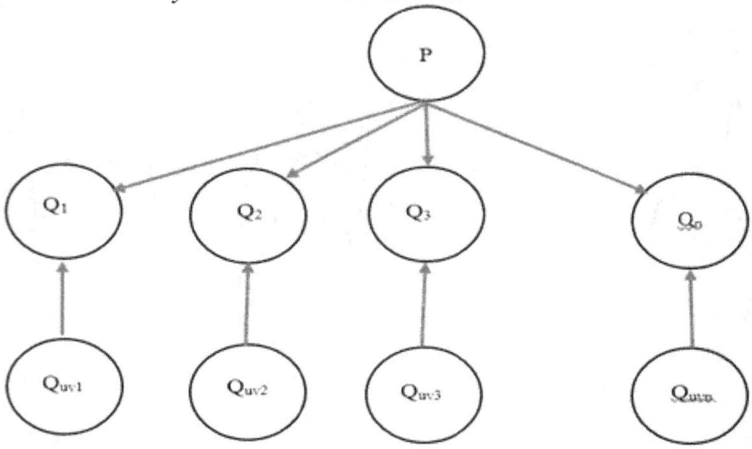

For text classification problems, naive Bayes classifiers work very well. They are capable of handling high-dimensional textual data, such the word vocabulary used in Bitcoin-related conversations, with effectiveness (Bustani, Mulyanto, & Nurjanana, 2018). Naive Bayes can withstand irrelevant characteristics, which frequently arise in noisy text data, because of the independence requirement. Figure 4 shows the comparison of real and forecast prices by the Naive Bayes model.

Figure 4. Assessment of real and predicted prices by Naive Bayes model

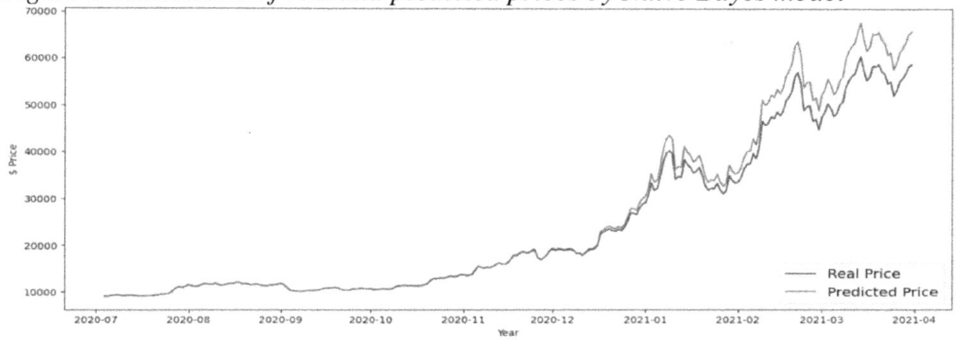

B. VADER

VADER is made expressly to assess sentiment in social media writing, it works particularly well with tweets, which frequently include slang, emoticons, and acronyms. Accurate emotion scores for brief, casual material is produced by combining lexical characteristics and grammatical rules. Four scores are generated by VADER sentiment analysis: the percentage of positive sentiment, the percentage of negative sentiment, the percentage of neutral sentiment, and a compound, which is a normalized, weighted composite score with a range of -1 (most negative) to +1 (most positive) (Weng, Liu, & Xiao, 2019). If there are multiple texts t_i at time t, aggregate the sentiment scores. Equations (3) to Equation (5) are the basic VADER model formulation.

$$VADER(t) \rightarrow \left(S_{pos}, S_{neg}, S_{neu}, S_{compound}\right) \quad (3)$$

$$S_{compound,t} = \frac{1}{n}\sum_{i=1}^{n} S_{compound,t_i} \quad (4)$$

For each time step t, creating a feature x_t:

$$x_t = \left[P_{t-1}, P_{t-2}, \ldots, V_{t-1}, V_{t-2}, \ldots, S_{compound,\, t}, S_{compound, t-1}, \ldots\right] \quad (5)$$

Where P_{t-i} are past Bitcoin prices and V_{t-i} are trading volumes. Since social media postings and tweets are frequently excellent sources of sentiment data for Bitcoin, VADER is specially designed to function effectively there (Pano & Kashef, 2020). More precise sentiment scores are provided by the way it manages social media. Figure 5 shows the assessment of real and predicted prices by VADER model.

Figure 5. Comparison of real and predicted prices by VADER model

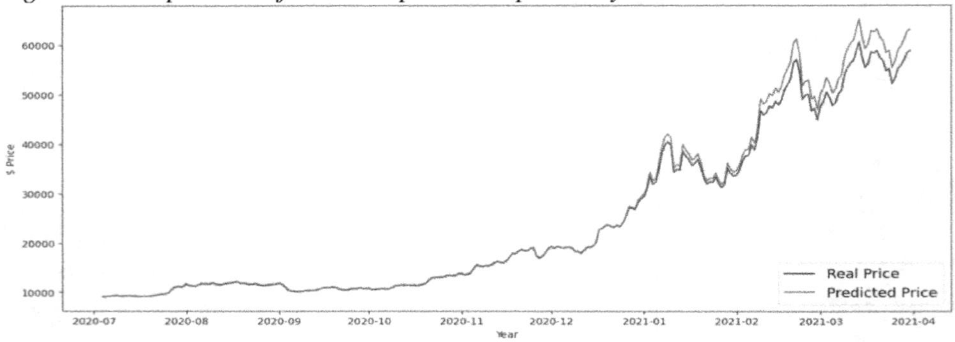

C. GRU

GRUs train more quickly than LSTMs because they handle sequential data more effectively and use less computing power. GRU models are ideal for assessing sentiment data that changes over time because of their ability to handle sequential data efficiently (Bin, Yang, Shen, Xie, Shen, & Li, 2019). Their ability to discern temporal correlations and patterns within the data is crucial in comprehending how attitudes shift and impact the value of Bitcoin. Sentiment data may be chaotic and loud, particularly when it comes from social media. GRUs can handle this kind of noisy input; they can learn to weed out unnecessary information and concentrate on the sentiment changes at the core that are important for understanding market dynamics (Tripathy, Nayak, & Prusty, n.d.). Equations (6) to Equation (9) are the basic GRU structure. Long-term dependencies in the text are captured by GRUs, which is essential for contextually comprehending sentiment.

$$r_t = \sigma\left(W_r \cdot [h_{t-1}, x_t] + b_r\right) \quad (6)$$

$$z_t = \sigma\left(W_z \cdot [h_{t-1}, x_t] + b_z\right) \quad (7)$$

$$h_t - \tanh\left(W_h \cdot \left[r_t \cdot h_{t-1}, x_t\right] + b_h\right) \tag{8}$$

$$h_t - \left(1 - z_t\right) \cdot h_{t-1} + z_t \cdot h_t \tag{9}$$

Where r_t is the reset gate at time t, σ is the sigmoid activation function, W_r is the weight matrix, b_r is the bias vector, h_{t-1} is the hidden state from the previous time step and x_t is the input at time t. The purpose of GRUs is to extract and utilize the temporal relationships present in sequential data. This enables the model to comprehend how historical feelings and price fluctuations affect present and future prices in the context of Bitcoin sentiment prediction (Zhao, Chetty, & Tran, 2019). In the financial markets, the sequence of events is important. Sequential processing of data by GRUs guarantees the preservation and efficient use of the chronological order of feelings and pricing (Tripathy, Hota, Mishra, Satapathy, & Nayak, 2024). Figure 6 shows the comparison of real and forecast prices by the GRU model.

Figure 6. Comparison of real and predicted prices by GRU model

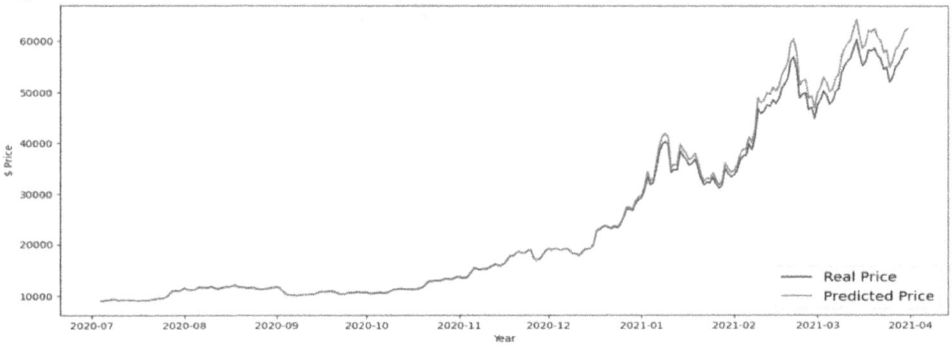

D. LSTM

LSTMs are useful for identifying the sentiment in lengthier tweets or a series of tweets since they are skilled at managing lengthy text sequences and keeping context over a prolonged amount of time (Wang, Cao, Yuan, & Cheng, 2021). To create forecasts for the upcoming time step or several time steps ahead, provide the model with past data (Fan, Wang, Gang, & Li, 2019). Equation (10) to Equation (14) shows the mathematical representation of LSTM structure.

$$i_t = \sigma\left(Y_i[p_{t-1}, x_t] + z_i\right) \tag{10}$$

$$f_t = \sigma\left(Y_f[p_{t-1}, x_t] + z_t\right) \tag{11}$$

$$c_t = f_t \times c_{t-1} + i_t \times tanp\left(Y_c[p_{t-1}, x_t] + z_c\right) \tag{12}$$

$$o_t = \sigma\left(Y_o[p_{t-1}, x_t] + z_o\right) \tag{13}$$

$$h_t = o_t \times \text{tanp}\left(c_t\right) \tag{14}$$

In the context where participation at the time step t is denoted by z_t, the hidden state by p_t, the cell state by c_t . Long-term knowledge retention is facilitated by a memory cell found in the LSTM design. Considering that the effects of noteworthy occurrences and feelings might last for several days or weeks, this is essential for predicting Bitcoin's mood (Tripathy, Balabantaray, Parida, & Nayak, 2024). LSTMs are resilient to noisy and erratic financial data because they can dynamically update their memory. As information gathered from social media and news sources may be quite inconsistent and noisy, this is advantageous for sentiment research of Bitcoin (Idrees, Alam, & Agarwal, 2019). Figure 7 shows the comparison of real and predicted prices by LSTM model.

Figure 7. Comparison of real and predicted prices by LSTM model

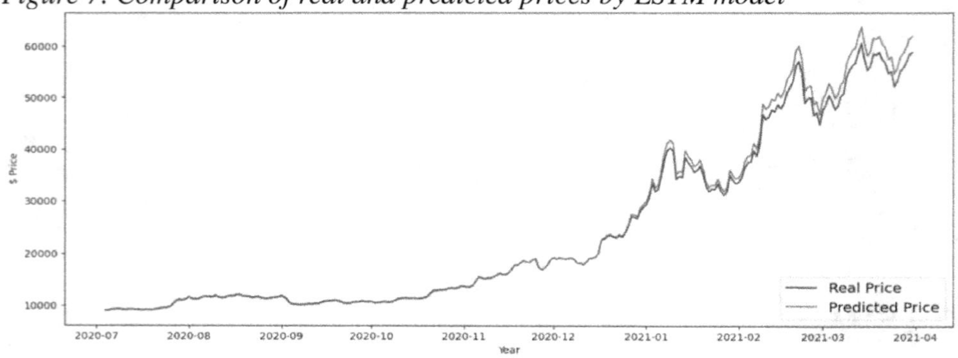

4. RESULT ANALYSIS

Investors obtain a more thorough knowledge of the sentiment of the market toward Bitcoin by combining sentiment research with technical indicators and news data (Bukhari, Raja, Sulaiman, Islam, Shoaib, & Kumam, 2020). In the dynamic and unstable cryptocurrency market, this all-encompassing strategy facilitates risk management, well-informed decision-making, and the creation of winning trading methods. Figure 8 shows the average daily sentiment analysis of Bitcoin price.

Figure 8. Average daily sentiment analysis plot

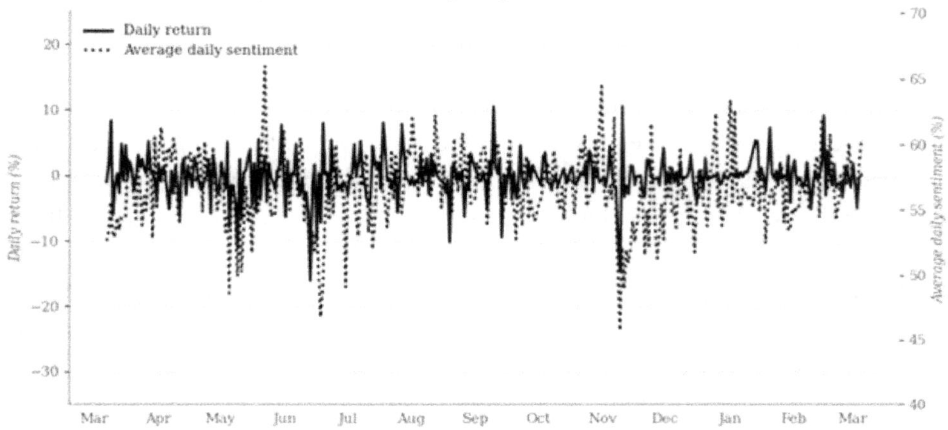

In sentiment analysis, Naive Bayes is frequently a useful baseline model. Its performance acts as a benchmark for comparing more intricate models, which aids in determining the extra benefit of sophisticated methods. VADER model is efficient in terms of computation it can analyse vast amounts of text data rapidly. For real-time sentiment analysis, where prompt insights might influence trading choices, this efficiency is essential. Table 1 shows the model accuracy, precision, recall and F1 score. The accuracy of the LSTM model is 0.952 which is minimal compared to other models. Figure 9 shows the bar diagram of the Accuracy, Precision, Recall and F1 scores of different models. GRUs can grasp how previous feelings impact present and future attitudes since they can retain knowledge for extended periods. In sentiment analysis, where past attitudes might have enduring consequences on market behaviour, this feature is essential. Sequences of data may be processed and used to teach LSTMs new things. Timing-sequenced text data from news stories and social media platforms is frequently used in Bitcoin sentiment research (Wang, He, & Liu, 2019). Accurate sentiment analysis depends on LSTMs' ability to recognize the temporal correlations and patterns in this data.

Table 1. Performance assessment of different models

Model Name	Accuracy	Precision	Recall	F1 Score
Naive Bayes	0.780	0.782	0.853	0.784
VADER	0.852	0.861	0.883	0.859
GRU	0.914	0.924	0.942	0.916
LSTM	0.952	0.937	0.961	0.954

Price changes can frequently be preceded by positive or negative sentiments. Investors can forecast market movements and make well-informed judgments using sentiment analysis. Sentiment analysis is useful in the evolution of public opinion regarding Bitcoin over time, which is important for developing long-term investment plans. Sentiment analysis is employed to direct policy choices by regulators by assisting them in understanding how the public views Bitcoin and identifying possible areas of enthusiasm or worry. A vital tool for all parties involved in the cryptocurrency ecosystem, sentiment analysis on Bitcoin may help with regulatory compliance, trading strategy improvement, market prediction, and business decision-making.

Figure 9. Bar diagram showing the Accuracy, Precision, Recall and F1 score of different models

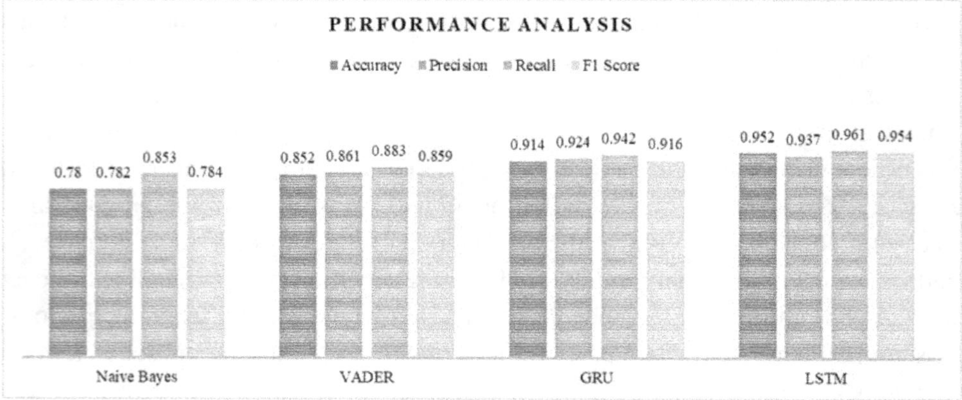

5. CONCLUSION

The market for cryptocurrencies has expanded quickly, and consumer knowledge of them is still rising. Compared to other models, the LSTM model behaves well and has a higher accuracy of 0.952. Sentiment research for Bitcoin frequently entails identifying intricate, non-linear correlations over time between different data points or among words in a phrase. LSTMs are excellent at simulating these complex patterns because of their advanced design. A wide range of input data sources, including text from news articles, tweets, forum posts, and even numerical data like past Bitcoin values, may be employed with LSTMs. Because of its adaptability, extensive sentiment analysis is made possible by combining various data sources.

As more institutions and private investors join the digital currency market, cryptocurrencies are becoming acknowledged as a kind of asset. Estimating the value of Bitcoin is, thus, an important financial undertaking. Our study shown that integrating economic indicators with the emotion index may be used to predict trends in the price of Bitcoin. Consequently, the input attributes utilized in further studies on the prediction of Bitcoin prices. There are issues with the current study that need to be fixed in the future. First, lexicon-based unsupervised learning performs less accurately than labelled data. Furthermore, lexicon-based learning might not perform as well with unstructured text input. Second, the trend was approximated every day instead of using inputs in the form of hours or minutes in real-time. In the end, the LSTM model was quite successful in predicting changes in price.

REFERENCE

Arslan, S. (2024, May 28). Bitcoin Price Prediction Using Sentiment Analysis and Empirical Mode Decomposition. *Computational Economics*, 1–22.

Bin, Y., Yang, Y., Shen, F., Xie, N., Shen, H. T., & Li, X. (2019, July). Describing video with attention-based bidirectional LSTM. *IEEE Transactions on Cybernetics*, *49*(7), 2631–2641. DOI: 10.1109/TCYB.2018.2831447 PMID: 29993730

Bukhari, A. H., Raja, M. A. Z., Sulaiman, M., Islam, S., Shoaib, M., & Kumam, P. (2020, April). Fractional neuro-sequential ARFIMA-LSTM for financial market forecasting. *IEEE Access : Practical Innovations, Open Solutions*, *8*, 71326–71338. DOI: 10.1109/ACCESS.2020.2985763

Bustani, P., Mulyanto, S., & Nurjanana (2018). Prediction of electricity sales using neural based inverse distance weighting method. *International Journal of Engineering & Technology, 7*(2.2), 65-69.

Dong, G., Li, R., Jiang, J., Wu, H., & McClure, S. C. (2020, September). Multigranular wavelet decomposition-based support vector regression and moving average method for service-time prediction on web map service platforms. *IEEE Systems Journal, 14*(3), 3653–3664. DOI: 10.1109/JSYST.2019.2944527

Doroslovački, K., Gradojevic, N., & Tarnaud, A. C. (2024, June 11). A Novel Market Sentiment Analysis Model for Forecasting Stock and Cryptocurrency Returns. *IEEE Transactions on Systems, Man, and Cybernetics. Systems*, *54*(9), 5248–5259. DOI: 10.1109/TSMC.2024.3402160

Fan, C., Wang, J., Gang, W., & Li, S. (2019). Assessment of deep recurrent neural network -based strategies for short-term building energy predictions. *Applied Energy*, *236*, 700–710. DOI: 10.1016/j.apenergy.2018.12.004

Feng, M., Zheng, J., Ren, J., Hussain, A., Li, X., Xi, Y., & Liu, Q. (2019, July). Big data analytics and mining for effective visualization and trends forecasting of crime data. *IEEE Access : Practical Innovations, Open Solutions*, *7*, 106111–106123. DOI: 10.1109/ACCESS.2019.2930410

Idrees, S. M., Alam, M. A., & Agarwal, P.Sheikh Mohammad Idrees. (2019). M Afshar Alam and Parul Agarwal, "A prediction approach for stock market volatility based on time series data". *IEEE Access : Practical Innovations, Open Solutions*, *7*, 17287–17298. DOI: 10.1109/ACCESS.2019.2895252

Jung, H. S., Lee, S. H., Lee, H., & Kim, J. H. (2023, August 1). Predicting Bitcoin Trends Through Machine Learning Using Sentiment Analysis with Technical Indicators. *Computer Systems Science and Engineering*, *46*(2), 2231–2246. DOI: 10.32604/csse.2023.034466

Liu, F., Cai, M., Wang, L., & Lu, Y. (2019, February). An ensemble model based on adaptive noise reducer and over-fitting prevention LSTM for multivariate time series forecasting. *IEEE Access : Practical Innovations, Open Solutions*, *7*, 26102–26115. DOI: 10.1109/ACCESS.2019.2900371

Pano, T., & Kashef, R. (2020). A Corpus of BTC Tweets in the Era of COVID-19. 2020 IEEE International IOT, Electronics and Mechatronics Conference (IEM-TRONICS) (pp. 1-4), IEEE. DOI: 10.1109/IEMTRONICS51293.2020.9216427

Ramani, K., Jahnavi, M., Reddy, P. J., VenkataChakravarthi, P., Meghanath, P., & Imran, S. K. (2023). Prediction of bitcoin price through LSTM, ARIMA, XGBoost, prophet and sentiment analysis on dynamic streaming data. 2023 9th International Conference on Advanced Computing and Communication Systems (ICACCS) (Vol. 1, pp. 1514-1518), IEEE.

Saleem, T., Yaqub, U., & Zaman, S. (2024, February 19). Twitter sentiment analysis and bitcoin price forecasting: Implications for financial risk management. *The Journal of Risk Finance*, *25*(3), 407–421. DOI: 10.1108/JRF-09-2023-0241

Tripathy, N., Balabantaray, S. K., Parida, S., & Nayak, S. K. (2024, June). Cryptocurrency fraud detection through classification techniques [IJECE]. *International Journal of Electrical and Computer Engineering*, *14*(3), 2918–2926. DOI: 10.11591/ijece.v14i3.pp2918-2926

Tripathy, N., Hota, S., & Mishra, D. (2023, September). Performance analysis of bitcoin forecasting using deep learning techniques. *Indonesian Journal of Electrical Engineering and Computer Science*, *31*(3), 1515–1522. DOI: 10.11591/ijeecs.v31.i3.pp1515-1522

Tripathy, N., Hota, S., & Mishra, D., Satapathy, P., & Nayak, S. K. (2024). Empirical Forecasting Analysis of Bitcoin Prices: A Comparison of Machine learning, Deep learning, and Ensemble learning Models. *International journal of electrical and computer engineering systems, 15*(1), 21-9. doi.org/10.32985/ijeces.15.1.3

Tripathy, N., Hota, S., Prusty, S., & Nayak, S. K. (2023). Performance Analysis of Deep Learning Techniques for Time Series Forecasting", In*2023 International Conference in Advances in Power, Signal, and Information Technology (APSIT)* (pp. 639-644). IEEE. DOI: 10.1109/APSIT58554.2023.10201734

. N. Tripathy, SK. Nayak, S. Prusty, (n.d.). A Comparative Analysis of Silverkite and Inter-Dependent Deep Learning Models for Bitcoin Price Prediction. Frontiers in Blockchain;7:1346410.

Wang, D., He, H., & Liu, D. (2019, July). Intelligent optimal control with critic learning for a nonlinear overhead crane system. *IEEE Transactions on Industrial Informatics*, *14*(7), 2932–2940. DOI: 10.1109/TII.2017.2771256

Wang, J., Cao, J., Yuan, S., & Cheng, M. (2021). Short-term forecasting of natural gas prices by using a novel hybrid method based on a combination of the CEEMDAN-SE-and the PSO-ALS-optimized GRU network. *Energy*, *233*(121082), 121082. DOI: 10.1016/j.energy.2021.121082

Weng, T., Liu, W., & Xiao, J. (2019, September). W. Liu and and J. Xiao, "Supply chain sales forecasting based on lightGBM and LSTM combination model". *Industrial Management & Data Systems*, *120*(2), 265–279. DOI: 10.1108/IMDS-03-2019-0170

Zhao, Y., Chetty, G., & Tran, D. (2019). Deep learning with XGBoost for real estate appraisal. Proc. IEEE Symp. Ser. Comput. Intell., pp. 1396-1401. DOI: 10.1109/SSCI44817.2019.9002790

Chapter 6
Integrating Green Information Systems as Organizational Performance Tools for the Mexican Manufacturing Industry

Enrique Ismael Melendez Ruiz
https://orcid.org/0000-0002-9624-6675
Universidad La Salle Victoria, Mexico

Demian Abrego-Almazan
https://orcid.org/0000-0003-0147-8834
Autonomous University of Tamaulipas, Mexico

Maria Ines Salas Rubio
https://orcid.org/0000-0002-2448-4308
Autonomous University of Tamaulipas, Mexico

ABSTRACT

This study examines the role of Green Information Systems (GIS) in shaping environmental strategies and organizational performance in Mexican manufacturing firms. GIS enhance efficiency, reduce costs, and mitigate environmental impact, yet their integration faces challenges due to complexity and workflow alignment issues. Using 154 survey responses analyzed with PLS-SEM, findings reveal that GIS significantly contribute to sustainability and competitiveness by advancing environmental

DOI: 10.4018/979-8-3693-8643-9.ch006

goals while improving economic performance. This study bridges a key research gap by providing empirical evidence on GIS adoption in Mexico, offering insights for businesses on leveraging GIS to achieve both ecological and financial benefits.

1. INTRODUCTION

The environment has been deteriorating at an alarming rate, posing a significant threat to both the planet and humanity, and thus becoming an urgent global concern (Pooja, 2024). This degradation is largely attributed to poor environmental practices by companies, including excessive resource consumption, energy inefficiency, and high levels of pollution (Watson & Boudreau, 2010). To address these issues, organizations have increasingly relied on tools such as information systems (IS) to optimize business operations, explore sustainable alternatives, and develop strategies that generate competitive advantages while maintaining environmental objectives (Brundtland, 1987; Molla et al., 2011; Sarkar & Young, 2009).

Research into the development of business strategies supported by information technologies (IT) to meet these objectives has significantly increased, as IT is considered essential for modern business operations (Meacham et al., 2013; Melville, 2013; Norheim, 2018). Moreover, IT serves as a crucial enabler for developing and implementing environmental practices, fostering business growth, and simultaneously mitigating environmental impact (Danso et al., 2019; Lo & Liao, 2021; Seidel et al., 2017).

In this context, the emergence of green information systems (GIS) has provided companies with the ability to enhance performance while reducing environmental impact without compromising economic goals (Baggia et al., 2019; Green et al., 2012a, Green et al., 2012b; Lembcke et al., 2021; Yang et al., 2018). However, achieving synergy between operational processes, the integration of GIS, and the development of environmental strategies heavily depends on the effective alignment of IT with organizational processes. This alignment poses a significant challenge due to the common absence of strategic planning, which is necessary to realize the projected benefits of such initiatives (Green et al., 2012a, Green et al., 2012b; Yang et al., 2018; Gorla et al., 2010; Ryoo & Koo, 2013).

To address these challenges, critical success factors (CSFs) have emerged as pivotal strategic elements. These include specific areas, conditions, or characteristics that require strong leadership from top management to ensure the successful consolidation of technology projects and their integration into organizational processes. Without such leadership, the adoption of technologies risks failing to deliver the expected benefits, which often justifies their substantial costs (Hwang & Min, 2015; Lei et al., 2023; Saade & Nijher, 2016; Petter et al., 2013; Talib et al., 2015).

In Mexico, there is a notable scarcity of studies addressing CSFs, GIS, and their synergistic effects with environmental strategies on organizational performance. Consequently, this study aims to analyze the critical success factors influencing green information systems and their subsequent impact on environmental strategies and organizational performance within medium and large companies in the manufacturing sector. The paper is structured as follows: first, the relevance of the topic is established; next, the theoretical framework and hypotheses are presented, followed by the methodology and results of the data analysis. Finally, the conclusions and limitations are discussed.

2. LITERATURE REVIEW

Information Systems (IS) are defined as IT tools designed to assist and support an organization's administrative and production processes, enabling the development of strategic plans to generate and sustain competitive advantages (Löser, 2015). The implementation of IS can significantly enhance business performance by improving the structure of operational processes, fostering innovation, and adopting technologies that positively impact the environment through the reduction of the organizational environmental footprint (Watson & Boudreau, 2010; Rajan & Baral, 2015). However, as with any project, the successful integration of IS requires meticulous planning and oversight by top management, as inadequate implementation can undermine the expected benefits (Gorla et al., 2010; Talib et al., 2015). To address this, various studies have identified a range of characteristics and strategic areas critical to achieving the performance and competitiveness objectives derived from IS value perception (Alhassan et al., 2019; Denolf et al., 201; Hu et al., 2018; Li et al., 2020). These elements are referred to as Critical Success Factors (CSFs) and are defined as a limited set of strategic areas that, when adequately addressed, ensure the successful completion and performance of technology projects (Ram & Corkindale, 2014). CSFs provide a foundation for strategy formulation, planning, project management, monitoring, and achieving organizational goals (Bullen & Rockart, 1981; Daniel, 1961; Li et al., 2016).

The integration of IS projects is a disruptive process that demands significant resource investment. In the absence of appropriate attention to CSFs, many technology projects achieve only limited success, as they fail to address the full project lifecycle and identify deficiencies that compromise their objectives (Garg & Agarwal, 2014; Reitsma & Hilletofth, 2018). Recognizing this challenge, numerous studies have explored how CSFs impact the successful integration of IS into organizational and operational processes (Li et al., 2020; Clark et al., 2020; Fonseca et al., 2019; Mahraz et al., 2020; Saygili et al., 2017). For this study, four key CSFs have been identified. The first is Process Reengineering Factors (PRFs), which involve redesigning organizational processes to ensure optimal compatibility with information systems, facilitating their effective integration and use (Eikebrokk et al., 2017). The second is Technology Factors (TFs), which refer to the availability and suitability of technological resources required for acquiring and implementing IS that align with organizational needs and objectives (Reitsma & Hilletofth, 2018). The third factor, Project Management Factors (PMFs), encompasses the composition, skills, and expertise of the project team, which are crucial for consolidating IS projects and ensuring their successful implementation (Yohannes & Mauritsius, 2022). Lastly, User Training Factors (UFs) emphasize equipping end users with the necessary knowledge and skills to maximize the benefits of IS and ensure their proper adoption and functionality (Garg & Agarwal, 2014).

Several studies have highlighted the significance of these factors in IS implementation. For example, managerial support, process reengineering, and project lifecycle monitoring have been shown to significantly influence the performance of manufacturing companies (Saygili et al., 2017; Chokalingam & Ramayah, 2013). From the user perspective, the composition of the work team, training, technical support, and system installation are among the most critical elements for IS success, as they ensure seamless information flow and a substantial impact on overall performance (Reitsma & Hilletofth, 2018). This underscores the importance of CSFs for the acquisition and implementation of green-focused IS. By addressing these elements, senior management can ensure that the significant investment in IS translates into measurable organizational benefits. Based on this, the following research hypothesis is proposed:

> H1: Critical success factors have a positive and significant influence on the performance of information systems in Mexican manufacturing companies.

In response to increasing environmental regulations, companies are adopting green initiatives to comply with these requirements and develop strategies for environmental sustainability (Garg & Agarwal, 2014; Butler, 2011; Kalyar & Shoukat, 2019). This has led to the integration of information systems (IS) as essential tools for enabling the continuous flow of information needed to support such strategies (Meacham et al., 2013). Companies have turned to Green Information Systems (GIS) as powerful instruments for achieving environmental objectives and reducing their environmental footprint (Baggia et al., 2019; Lembcke et al., 2021; Kalyar & Shoukat, 2019; Gholami et al., 2018).

GIS consists of three fundamental elements, as defined by Hart (1997). The first element, Pollution Prevention (PP), refers to innovations and the utilization of IS to minimize pollutant emissions, thereby addressing environmental challenges proactively (Green et al., 2012a; Lembcke et al., 2021; Khan & Qianli, 2017). The second, Product Stewardship (PS), encompasses management practices aimed at reducing the environmental impact associated with a product's life cycle. These practices include recycling, green manufacturing, and the design of eco-friendly products, ensuring sustainability throughout the production and consumption phases. Finally, the third element, Sustainable Development (SD), is defined as the long-term growth of a company through the adoption and integration of clean technologies, which enable organizations to align their operations with environmental and economic objectives (Li et al., 2016; Martínez-Arroyo & Valenzo-Jiménez, 2022). Together, these elements provide a comprehensive framework for GIS, enabling companies to implement strategies that balance operational efficiency with environmental sustainability. GIS have been widely recognized in literature as tools capable of partially addressing environmental challenges by enabling companies to reduce their ecological footprint. They support processes such as resource reuse, energy optimization, logistics, and carbon emissions management (Abbas et al, 2022; Khor et al., 2015). Beyond pollution reduction, GIS also influences organizational performance. Studies such as those by Khan & Qianli (2017) and Anthony (2019) demonstrate a direct or indirect positive relationship between GIS and organizational performance (RORG), particularly within the context of green supply chains in manufacturing companies. These findings challenge the notion that adopting green business practices lacks financial benefits.

Furthermore, research has identified a direct and positive impact of GIS on the environmental and financial performance of small and medium enterprises (SMEs), with medium- and long-term benefits (Baggia et al., 2019; Lembcke et al., 2021; Yang et al., 2017). These technologies not only prevent pollution but also optimize resource utilization and reduce operational costs. By implementing GIS, companies can significantly improve their processes while contributing to environmental sustainability. This dual benefit underscores the strategic value of GIS in achieving both environmental and organizational objectives. Based on this evidence, the following research hypothesis is proposed:

> H2: Green information systems significantly and positively impact the organizational performance of Mexican manufacturing companies.

Environmental sustainability has become one of the most pressing social concerns (Mclean & Wilson, 2011), particularly because manufacturing companies are significant contributors to environmental degradation. However, their resources, capabilities, and knowledge also position them as key agents of change toward cleaner development practices (Abbas et al., 2022). This dual role has driven the interest of companies in creating strategies that incorporate the development and implementation of environmental systems (Loeser et al., 2017). Uncoordinated or incoherent initiatives that fail to align with process transformations tend to be inefficient (Suša et al., 2020). Consequently, formulating robust environmental strategies is fundamental for achieving corporate sustainability (Danso et al., 2019). For the purposes of this study, environmental strategies are defined as the mutual understanding between top management and IT executives regarding future opportunities and challenges related to environmental sustainability and long-term company growth (Loeser et al., 2017). These strategies represent the company's commitment to meeting environmental goals while enhancing the efficiency and effectiveness of operations supported by GIS (Danso et al., 2019; Loeser et al., 2017; Chen et al., 2018; Wu & Wu, 2014).

Despite their potential, the nature and benefits of these initiatives are often not well understood. This creates challenges for management in integrating technologies with environmental care strategies (Loeser et al., 2017). This gap has motivated studies analyzing the synergy between GIS and environmental processes aimed at minimizing environmental footprints while leveraging opportunities to gain a competitive advantage. For example, Anthony (2019) developed a model to identify factors influencing the use of GIS in ecological practices, finding a significant relationship between environmental strategies and IS implementation. Additionally, other studies (Danso et al., 2019; Green et al., 2012b; Li et al., 2016) argue that while environmental strategies alone may not guarantee a long-term competitive advantage, integrating GIS into green supply chains and designing eco-friendly products allows companies to achieve this goal. These technologies play a critical role in fulfilling green objectives by supporting environmental product design and optimizing resource management. Although limited research has explored how information technologies support environmental sustainability strategies, existing studies highlight the synergy between these elements. Having the technological infrastructure to support the activities demanded by these strategies ensures that they can be executed effectively and on time. Based on this evidence, the following hypothesis is proposed:

> H3: The use of green information systems significantly and positively influences the environmental strategies of Mexican manufacturing companies.

For this study, organizational performance is defined as a firm's ability to achieve its objectives and goals by utilizing resources effectively and efficiently. It comprises three key elements: operational performance (OP), referring to the firm's capacity to produce and deliver products and services to customers in a professional and efficient manner (Masa'deh et al., 2017); financial performance (FP), focusing on return on investment and profitability (Danso et al., 2019; Green et al., 2012a; Zhang

& Wei, 2021); and environmental performance (EP), which measures a firm's ability to reduce air pollutant emissions, effluents, and solid waste, as well as to minimize the use of toxic and hazardous resources and materials (Hernández & Nishii, 2023; Green et al., 2012b). Empirical studies have extensively analyzed the influence of environmental strategies on manufacturing companies worldwide. For example, Li et al. (2016) and Acuña et al. (2017) demonstrated that active and functional strategies in green supply chains strongly and positively influence both environmental and financial performance over the long term, highlighting the profitability of integrating robust information systems into business operations.

Environmental strategies (ESTVs) also play a crucial role in enhancing the integration and functionality of GIS. When combined with activities that promote awareness of environmental challenges and encourage eco-friendly practices, these strategies contribute not only to improved environmental performance but also to overall organizational performance. This is achieved by fostering better integration and efficiency in ongoing organizational processes (Baggia et al., 2019; Ryoo & Koo, 2013; Yang et al., 2017; Loeser et al., 2017). Furthermore, when management views environmental sustainability as a strategic opportunity, it can develop strategies that integrate environmental practices into core business processes (Loeser et al., 2017). However, uncoordinated environmental initiatives that lack clear strategies, guidelines, and policies are unlikely to succeed significantly (Bansal & Agarwal, 2015). Therefore, based on empirical evidence, the following hypothesis is proposed:

> H4: Environmental strategies significantly and positively influence the organizational performance of Mexican manufacturing firms.

In summary, this study explores the interplay between Critical Success Factors (CSFs), Green Information Systems (GIS), and environmental strategies in shaping organizational performance. By examining these relationships, the research underscores the strategic importance of integrating sustainability into information systems and broader business practices. The findings aim to provide actionable insights for achieving both competitive advantage and sustainable development within the Mexican manufacturing sector.

3. METHOD

The present study employs a causal, quantitative, and non-experimental research approach with an exploratory cross-sectional design to achieve the proposed objectives. The procedure began with a comprehensive review of specialized literature, which served as the foundation for the development of the hypotheses. Subsequently, a list of items for the research instrument was generated. These items were validated by researchers and experts in the field, whose feedback and suggestions were incorporated into the final version of the instrument. The resulting questionnaire comprised 7 general questions and 50 items, measured on a 7-point Likert scale, where 1 corresponds to "totally disagree" and 7 to "totally agree."

Data collection was conducted through a non-probabilistic convenience sampling method during the first quarter of 2023. While the required sample size was calculated to be 286 middle managers (with a 95% confidence level), the final sample consisted of 158 participants, representing 58% of the originally required sample. To address this limitation, the statistical software GPower 3.1 was used to verify the adequacy of the collected sample for detecting medium effect sizes in the proposed causal relationships (Faul et al., 2009). The parameters used for this verification included a mean effect size $f^2 = 0.15$, a significance level (α) of 0.05, a statistical power of 95%, and seven predictors. The results confirmed that the sample was sufficient to determine the influence and statistical significance of the relationships under study. Finally, the collected data were analyzed using the Structural Equation Modeling (SEM) method, specifically employing the Partial Least Squares (PLS) technique. This approach was chosen for its robustness in analyzing causal relationships within exploratory research contexts

4. RESULTS AND DISCUSSIONS

For the inferential analysis, the PLS technique indicates that the process must be carried out in two stages: i) validation of the measurement model and ii) validation of the structural model, but before these two stages, it is necessary to evaluate the model fit and thus detect discrepancies between the information evaluated by the model and the one predicted. The results obtained for this evaluation can be seen in table 1 which is within the range established by (Hair et al., 2017), allowing us to continue analyzing the model results.

Table 1. Fit indicators results.

Model fit indicators			
Coefficient	**Value**	**Confidence intervals**	
		95%	**99%**
SRMR	0.059	0.045	0.060
d_ULS	2.052	1.810	2.164
d_G	1.724	1.588	1.792

Source: elaboration

Continuing with the model evaluation process, the first stage comprising the measurement model analysis is carried out through the validation and evaluation of the instrument's reliability. First, the individual reliability test was performed to verify an indicator as part of a reflective construct through its factorial loading (λ), which must be greater than 0.707 (Hair et al., 2017). Considering this, the indicators FGP1; FPG2; FRP1; FRP2; FRP3; FRP3; FIT3; FU1; FU2; PP3; PS2; SD2; RA3; RA4; RF3; RF4; RO4 were excluded. Table 2 shows the items that were maintained; however, the indicators FPG4; FRP4; FRP4; FU3; PS4; and RF5 do not comply with this rule but were still considered because their presence did not negatively affect the measurement parameters of the model.

Table 2. Individual reliability

Item	Description	λ	Source
FGP3	Project management expertise ensures effective system acquisition.	0.815	(Garg & Agarwal, 2014; Reitsma & y Hilletofth; 2018)
FGP4	Defined goals, timelines, and metrics guide system acquisition and implementation.	0.660	
FGP5	Structured plans drive system acquisition, upgrades, and implementation.	0.787	
FRP4	Continuous evaluation improves system-supported processes.	0.694	(Garg & Agarwal, 2014; Lei et al., 2023)
FTI1	Established procedures for monitoring and acquisition of information systems.	0.730	(Eikebrokk & Iden, 2017; Reitsma & Hilletofth, 2018)
FTI2	A qualified project manager oversees system acquisition.	0.717	
FU3	Clear goals, deadlines, and metrics for system acquisition and implementation.	0.616	
PP1	Systems track and report environmental impacts of manufacturing processes.	0.716	(Anthony, 2019; Baggia et al., 2019; Ruiz et al., 2023; Yang et al., 2017, 2018)
PP2	Systems ensure compliance with environmental, health, and safety standards.	0.802	
PP4	Systems enable monitoring and control of polluting emissions.	0.798	
PS1	Systems support sustainable material procurement practices.	0.835	
PS3	Systems aid in the production of eco-friendly products.	0.781	
PS4	Systems enable production using recycled or leftover materials.	0.702	
PS5	Systems optimize delivery routes to reduce fuel consumption.	0.756	
SD1	Systems ensure adherence to environmental standards.	0.780	
SD3	Systems help identify and deactivate unused services to reduce waste.	0.760	
SD4	Systems streamline operations across all organizational departments.	0.791	
SD5	Systems to minimize paper usage in operations and communications.	0.715	
ESTV1	Ecological impact is prioritized in IT project selection.	0.889	(Chen et al., 2018; Löser, 2015; Molla et al., 2011; Ruiz et al., 2023)
ESTV2	Performance metrics assess the environmental impact of IT initiatives.	0.889	
ESTV3	Resources are allocated to support environmental strategies.	0.775	
ESTV4	Senior management acknowledges the benefits of environmental strategies.	0.851	
ESTV5	Senior management actively supports green initiatives.	0.801	
ESTV6	Senior management views systems as critical for achieving green goals.	0.781	
ESTV7	Senior management quickly identifies and acts on environmental opportunities.	0.800	

continued on following page

Table 2. Continued

Item	Description	λ	Source
RA1	Systems to enhance material reuse in manufacturing processes.	0.785	(Feng et al., 2018; Green et al., 2012a, 2012b; Ruiz et al., 2023; Wu & Wu, 2014; Yang et al., 2017, 2018, 2019).
RA2	Systems to reduce the use of toxic materials in production.	0.885	
RA5	Systems to lower energy consumption across operations.	0.837	
RF1	Systems do contribute to higher returns on sales.	0.742	
RF2	Systems drive improved returns on investment.	0.708	
RF5	Systems support growth in market share.	0.703	
RO1	Systems standardize processes and boost operational efficiency.	0.780	
RO2	Systems usage shorten new product development cycles.	0.790	
RO3	Systems enable seamless value chain integration.	0.728	

Source: adapted from Ruiz et al. (2023).

Once the reliability of the item has been verified, we proceed to evaluate the constructs' reliability by analyzing two internal consistency measures: the composite reliability coefficient (fc) and the rho_A. For both, we resort to the specifications of (Hair et al., 2017), who indicate that both should be greater than 0.7. Table 3 shows the results obtained and as can be seen; the constructs used to achieve adequate internal consistency.

Now, it is time to analyze the convergent and discriminant validity of the proposed constructs. The Average Variance Extracted (AVE) coefficient assesses the former, which should have a recommended minimum value of 0.5 (Fornell & Larcker, 1981). Regarding discriminant validity, its coefficients were calculated using the Heterotrait Monotrait Ratio of Correlations (HTMT) method, which indicates that the values obtained in the matrix should all be less than 0.85, which shows that the constructs analyzed are different from each other (Henseler et al., 2015), i.e., there are no discriminant validity problems, all the above mentioned is demonstrated in table 3.

Table 3. Reliability, convergent, and discriminant validity.

Variables	Reliability		Convergent Validity	Discriminant validity (HTMT criterion)				Variance explained
	rho_A	fc	AVE	ESTV	CSF	RORG	GIS	R^2
ESTV	0.940	0.938	0.685	--				0.718
CSF	0.886	0.882	0.518	0.712	--			--
RORG	0.933	0.931	0.601	0.703	0.664	--		0.580
GIS	0.941	0.940	0.590	0.846	0.740	0.749	--	0.551

Source: Own elaboration.

Once it has been verified that the proposed model satisfies the requirements of the measurement model, the structural model is analyzed by evaluating the predictive power (variance explained, R^2) and the strength (standardized path coefficients) of the proposed hypotheses. To assess the explained variance, the guide of (Hair et al., 2017) is taken; for this, R^2 values of 0.25; 0.50; and 0.75 for dependent constructs are considered as weak, moderate, and high, respectively. From this criterion (table 3), the R^2 of the proposed constructs are in moderate to high ranges. In the case of structural model analysis, the values obtained for the path coefficients are analyzed to estimate the influence of the independent variables in explaining the dependent variables. For this purpose, the Bootstrapping technique was used, where the 158 cases collected plus a set of 5000 subsamples were considered. Finally, table 4 and Figure 1 show the results of the structural model.

Table 4. Structural model results.

Hypothesis		β	*t* Values	*P* Values	Comment
H1	**CSF -> GIS**	**0.744**	**12.837**	**0.000**	**Accepted**
H2	GIS -> RORG	0.541	3.926	0.000	Accepted
H3	GIS -> ESTV	0.848	26.671	0.000	Accepted
H4	ESTV -> RORG	0.251	1.724	0.042	Accepted

Source: Own elaboration.

Figure 1. Proposed structural model results.

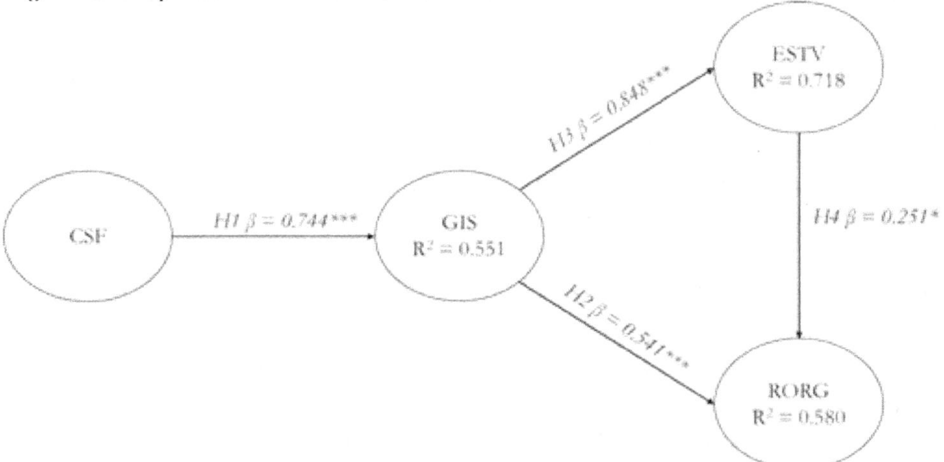

Source: Own elaboration.

5. DISCUSSIONS

As shown in Table 4, hypothesis H1 is supported, indicating that Critical Success Factors (CSFs) have a positive and significant influence ($\beta=0.744$, $p<0.001$) on the successful implementation of Green Information Systems (GIS) in Mexican manufacturing companies. This result highlights the importance of management, particularly top management, in allocating sufficient resources and providing appropriate attention to ensure the seamless integration of GIS into manufacturing processes. These findings are consistent with previous studies (Alhassan et al., 2019; Reitsma & Hilletofth, 2018; Yohannes & Mauritsius, 2022), as well as the works of Abrego et al. (2018) and Hwang & Min (2015), which analyzed the post-implementation phase of IS and emphasized the need to adhere to CSF guidelines to maintain a continuous positive impact on companies' economic and operational performance. In this regard, Pérez (2021) and García & Rodríguez (2022) have demonstrated that strategic planning, leadership, and staff training are decisive elements in the implementation of GIS, aligning with digital transformation and sustainability goals.

The results of this study further underscore the critical role of CSFs in enabling the effective implementation of GIS in the manufacturing sector. Among these factors, process reengineering is particularly significant, ensuring that GIS are aligned seamlessly with existing operational workflows. This alignment optimizes efficiency and minimizes potential disruptions during the integration process. Additionally, employee training is indispensable for fully leveraging the capabilities of GIS technologies, as it equips employees with the necessary skills and knowledge to use these systems effectively. Finally, consistent managerial support is a decisive factor in overcoming resistance to change, fostering a supportive environment that encourages adoption and ensures the successful integration of GIS into organizational processes. Collectively, these factors provide a strategic roadmap for companies seeking to implement GIS as part of their sustainability and performance enhancement strategies.

Regarding hypothesis H2, the results confirm a positive and significant relationship between GIS and organizational performance ($\beta=0.541$, $p<0.001$). This finding aligns with previous studies (Yang et al., 2018; Yang et al., 2019; Nanath & Pillai, 2017), as well as research by Qu & Liu (2022), which has shown that the implementation of GIS improves supply chain integration and fosters sustainable innovation, translating into a positive impact on operational efficiency and financial performance. GIS enable companies to reduce their environmental footprint, enhance control over manufacturing processes, and optimize resource utilization. These capabilities result in substantial improvements in operational, financial, and organizational domains. Additionally, research has shown that GIS can generate tangible economic benefits, such as cost reduction through efficient resource management and process optimization (Velásquez et al., 2019). Furthermore, these systems improve decision-making by providing real-time data and advanced analytics, enabling top management to respond swiftly to challenges and capitalize on opportunities. These findings emphasize the role of GIS as key enablers of sustainable growth and competitiveness in the manufacturing industry.

Relating hypothesis H3, the results show a strong and positive influence of GIS on environmental strategies ($\beta=0.848$, $p<0.001$). This relationship reflects the fundamental role of GIS in the development, execution, and monitoring of environmental strategies within organizations. In particular, GIS facilitate the integration of sustainable practices into organizational processes, such as the adoption of green supply chains, the implementation of eco-friendly production methods, and the design of environmentally responsible products. Purnomo et al. (2022) mention that the use of green information technologies not only enables efficient resource management and carbon emission reduction but also contributes to compliance with environmental regulations and enhances corporate reputation. Similarly, Qu & Liu (2022) found that the integration of GIS within the supply chain allows companies to adopt green innovations more effectively, optimizing processes and strengthening business sustainability. Moreover, GIS drive internal transformation within companies by aligning their technological capabilities with sustainability goals.

Finally, for hypothesis H4, the relationship between environmental strategies and organizational performance is confirmed ($\beta=0.251$, $p<0.05$). Quiroz y Vera (2023) demonstrate that corporate environmental ethics fosters green innovation, which positively impacts organizational performance. Lei (2024) has shown that optimizing supply chain networks aimed at reducing emissions improves operational efficiency and business competitiveness.

Research has demonstrated that the adoption of sustainable environmental practices and innovation is fundamental to improving organizational performance and achieving sustainable competitive advantage (Martínez-Arroyo et al., 2022). Similarly, the implementation of circular economy practices is positively correlated with improvements in organizational performance (Hernández & Nishii, 2023). The adoption of environmental management systems based on ISO 14001 can also enhance environmental and operational performance, positively influencing organizational performance (Gélvez & Serna, 2023). Finally, different types of environmental innovations have a positive impact on key organizational performance metrics, such as sales, employment, and productivity (Biggi, Mina & Tamagni, 2023)

6. CONCLUSIONS

The objective of this research was to determine how CSFs influence the integration of GIS and how these, in turn, affect environmental care strategies and organizational performance in medium and large companies within the Mexican manufacturing sector. The findings underscore the pivotal role of GIS in enhancing organizational performance by optimizing operational activities, streamlining product delivery, and improving employee efficiency. Additionally, GIS contributes to increasing customer satisfaction, reducing product life cycles, and generating indirect financial benefits such as higher average sales and cost savings through more efficient use of materials and resources. On the environmental side, GIS facilitate significant reductions in carbon emissions, minimize material waste, and provide greater control over the use of toxic materials. These benefits are maximized when GIS are implemented under CSF guidelines, which minimize risks associated with the integration of robust information systems, mitigate deficiencies during the integration process, and ensure the realization of projected benefits.

Moreover, the study highlights the lucrative potential of environmental care strategies when supported by information technology. This approach reduces uncertainties regarding the economic impacts of environmental protection initiatives by enabling companies to incorporate such strategies into their business processes. The active involvement of top management emerges as a critical factor, as it allows companies to respond promptly to early signals of environmental opportunities and transform them into competitive advantages. By doing so, companies not only reduce their environmental footprint but also position themselves as socially responsible and competitive players in the market. The elements analyzed in this study provide an empirical perspective on the impact of GIS in driving the development of the Mexican manufacturing sector. The findings contribute to the ongoing discourse in the literature on environmental management, business management, and information systems. They also help identify guidelines with the greatest influence on both environmental protection and organizational growth. This research offers a foundation for understanding how GIS can serve as a bridge between sustainability objectives and enhanced business performance.

Furthermore, the results are applicable to a wide range of companies, especially in the context of business management. By leveraging these insights, both managers and employees can better respond to pressures related to corporate performance. This capability allows the private sector to remain competitive, attract new customers, access new markets, and improve public and investor profiles. These advancements can translate into greater economic performance while contributing to environmental sustainability. Over time, this could foster a virtuous cycle within the country, promoting a shift toward a more socially responsible and sustainable corporate culture.

To address the practical implications of these findings, manufacturing companies in Mexico can adopt specific strategies to overcome challenges in the implementation of GIS. Research suggests that the implementation of environmental strategies through GIS can generate sustainable competitive advantages, facilitating compliance with environmental regulations and optimizing the use of organizational resources (Purnomo et al., 2022). The implementation of these measures will enable companies to successfully integrate GIS into their operations, simultaneously achieving sustainability and competitive advantage.

Limitations and Future Research

Despite the valuable insights offered, this study has certain limitations that must be acknowledged. The sample size and geographical scope were restricted, which could limit the generalizability of the findings to other sectors or regions. Additionally, the specific characteristics of the Mexican manufacturing industry may not fully represent other industries or geographical contexts. Future studies could expand this research to different industries, regions, or even small and medium-sized enterprises (SMEs), providing a broader understanding of GIS implementation and its impact across diverse organizational settings. Another challenge identified during this study was the training of personnel to use GIS effectively. Such training requires substantial time and resources, often posing a significant barrier to successful implementation. Additionally, internal resistance to change was frequently observed, a common obstacle in technological adoption projects that highlights the importance of fostering an organizational culture that embraces innovation and sustainability.

Future research could explore the adaptability and effectiveness of GIS in other industrial sectors or SMEs, broadening the understanding of their impact and challenges in different organizational contexts. Additionally, connecting these findings with global trends in sustainability and green technology would provide a comparative framework, emphasizing the relevance of GIS implementation on an international scale. Such research could further illuminate the role of GIS in driving sustainable and competitive practices globally.

REFERENCES

Abbas, A., Luo, X., Wattoo, M. U., & Hu, R. X. (2022). Organizational Behavior in Green Supply Chain Integration: Nexus Between Information Technology Capability, Green Innovation, and Organizational Performance. *Frontiers in Psychology*, *13*, 874639. Advance online publication. DOI: 10.3389/fpsyg.2022.874639 PMID: 35391968

Abrego, D., Sánchez-Gutiérrez, J., & Medina-Quintero, J. M. (2018). Influence of information systems on organizational results. *Contaduría y Administración*, *63*(1), 1–20. DOI: 10.22201/fca.24488410e.2018.1256

Acuña, N., Figueroa, L., & Wilches, M. J. (2017). Influencia de los Sistemas de Gestión Ambiental ISO 14001 en las organizaciones: Caso estudio empresas manufactureras de Barranquilla. *Ingeniare. Revista Chilena de Ingeniería*, *25*(1), 143–153. DOI: 10.4067/S0718-33052017000100143

Alhassan, I., Sammon, D., & Daly, M. (2019). Critical success factors for data governance: A telecommunications case study. *Journal of Decision Systems*, *28*(1), 41–61. DOI: 10.1080/12460125.2019.1633226

Anthony, B.Jr. (2019). Green information system integration for environmental performance in organizations: An extension of belief-action-outcome framework and natural resource-based view theory. *Benchmarking*, *26*(3), 1033–1062. DOI: 10.1108/BIJ-05-2018-0142

Baggia, A., Maletič, M., Žnidaršič, A., & Brezavšček, A. (2019). Drivers and outcomes of green IS adoption in small and medium-sized enterprises. *Sustainability (Basel)*, *11*(6), 1575. DOI: 10.3390/su11061575

Bansal, V., & Agarwal, A. (2015). Enterprise resource planning: Identifying relationships among critical success factors. *Business Process Management Journal*, *21*(6), 1337–1352. DOI: 10.1108/BPMJ-12-2014-0128

Biggi, G., Mina, A., & Tamagni, F. (2023). There are different shades of green: heterogeneous environmental innovations and their effects on firm performance. *arXiv preprint arXiv:2310.08353*.Disponible en: https://arxiv.org/abs/2310.08353

Brundtland, G. (1987). Our Common Future—Call for Action. *Environmental Conservation*, *14*(4), 291–294. DOI: 10.1017/S0376892900016805

Bullen, C. V., & Rockart, J. F. (1981). A primer on critical success factors. http://mit.dspace.org/bitstream/handle/1721.1/1988/SWP-1220-08368993- CISR-069.pdf?sequence=1

Butler, T. (2011). Compliance with institutional imperatives on environmental sustainability: Building theory on the role of Green IS. Journal of strategic information systems, 20(1, SI), 6–26. https://doi.org/DOI: 10.1016/j.jsis.2010.09.006

Chen, C. S., Yu, C. C., & Hu, J. S. (2018). Constructing performance measurement indicators to suggested corporate environmental responsibility framework. *Technological Forecasting and Social Change*, *135*, 33–43. DOI: 10.1016/j.techfore.2017.05.033

Clark, J. A., Liu, Y., & Isaias, P. (2020). Critical success factors for implementing learning analytics in higher education: A mixed-method inquiry. *Australasian Journal of Educational Technology*, *36*(6), 89–106. DOI: 10.14742/ajet.6164

Daniel, D. R. (1961). Management information crisis. *Harvard Business Review*, 111–121.

Danso, A., Adomako, S., Amankwah-Amoah, J., Owusu-Agyei, S., & Konadu, R. (2019). Environmental sustainability orientation, competitive strategy, and financial performance. *Business Strategy and the Environment*, *28*(5), 885–895. DOI: 10.1002/bse.2291

Denolf, J. M., Trienekens, J. H., Wognum, P. M., Van Der Vorst, J. G. A. J., & Omta, S. W. F. (2015). Towards a framework of critical success factors for implementing supply chain information systems. *Computers in Industry*, *68*, 16–26. DOI: 10.1016/j.compind.2014.12.012

Eikebrokk, T. R., & Iden, J. (2017). Strategising IT service management through ITIL implementation: Model and empirical test. *Total Quality Management & Business Excellence*, *28*(3-4), 238–265. DOI: 10.1080/14783363.2015.1075872

Faul, F., Erdfelder, E., Buchner, A., & Lang, A. G. (2009). Statistical power analyses using G* Power 3.1: Tests for correlation and regression analyses. *Behavior Research Methods*, *41*(4), 1149–1160. DOI: 10.3758/BRM.41.4.1149 PMID: 19897823

Feng, M., Yu, W., Wang, X., Wong, C. Y., Xu, M., & Xiao, Z. (2018). Green supply chain management and financial performance: The mediating roles of operational and environmental performance. *Business Strategy and the Environment*, *27*(7), 811–824. DOI: 10.1002/bse.2033

Fonseca, P. G., Souza, D. M. O. R., Santos, E. M., & Miranda, M. A. S. (2019). Fatores críticos de sucesso na implementação de sistemas de informação a partir da percepção de usuários: Uma experiência no setor público. *Holos*, *35*(3), 1–19. DOI: 10.15628/holos.2019.8190

Fornell, C., & Larcker, D. F. (1981). Structural equation models with unobservable variables and measurement error. Algebra and statistics. https://doi.org/DOI: 10.1177/002224378101800313

García, J. D., & Rodríguez, M. A. (2022). Factores críticos de éxito en la implementación de un ERP en una empresa del sector de alimentos. *Trabajo de grado, Universidad Nacional de Colombia.*https://repositorio.unal.edu.co/bitstream/handle/unal/81260/74130731.2022.pdf

Garg, P., & Agarwal, D. (2014). Critical success factors for ERP implementation in a Fortis hospital: An empirical investigation. *Journal of Enterprise Information Management*, *27*(4), 402–423. DOI: 10.1108/JEIM-06-2012-0027

Gholami, R., Molla, A., Goswami, S., & Brewster, C. (2018). Green information systems use in social enterprise: The case of a community-led eco-localization website in the West Midlands region of the UK. *Information Systems Frontiers*, *20*(6), 1345–1361. https://link.springer.com/article/10.1007/s10796-016-9733-z. DOI: 10.1007/s10796-016-9733-z

Gorla, N., Somers, T. M., & Wong, B. (2010). Organizational impact of system quality, information quality and service quality. *The Journal of Strategic Information Systems*, *19*(3), 207–228. DOI: 10.1016/j.jsis.2010.05.001

Green, K.Jr, Zelbst, P., Bhadauria, V., & Meacham, J. (2012b). Do environmental collaboration and monitoring enhance organizational performance? *Industrial Management & Data Systems*, *112*(2), 186–205. DOI: 10.1108/02635571211204254

Green, K.Jr, Zelbst, P. J., Meacham, J., & Bhadauria, V. S. (2012a). Green supply chain management practices: Impact on performance. *Supply Chain Management*, *17*(3), 290–305. DOI: 10.1108/13598541211227126

Hair, J. F., Hult, G. T. M., Ringle, C. M., & Sarstedt, M. (2017). *A Primer on Partial Least Squares Structural Equation Modeling (PLS-SEM)* (2nd ed.). Sage.

Hart, S. (1997). Beyond greening: Strategies for a sustainable world. *Harvard Business Review*, *75*(1), 66–77.

Henseler, J., Ringle, C. M., & Sarstedt, M. (2015). A new criterion for assessing discriminant validity in variance-based structural equation modeling. *Journal of the Academy of Marketing Science*, *43*(1), 115–135. DOI: 10.1007/s11747-014-0403-8

Hernández P., J., & Nishii, Y. (2021). Análisis de información y factores de desempeño ambiental y de economía circular en empresas peruanas. Comuni@cción, 12(1), 37-52. https://doi.org/DOI: 10.33595/2226-1478.12.1.481

Hu, Y. Y., Zhu, Y., Tucker, J., & Hu, Y. (2018). Ownership influence and CSR disclosure in China. *Accounting Research Journal*, *31*(1), 8–21. DOI: 10.1108/ARJ-01-2017-0011

Hwang, D., & Min, H. (2015). Identifying the drivers of enterprise resource planning and assessing its impacts on supply chain performances. *Industrial Management & Data Systems*, *115*(3), 541–569. DOI: 10.1108/IMDS-10-2014-0284

Kalyar, M. N., Shoukat, A., & Shafique, I. (2019). Enhancing firms' environmental performance and financial performance through green supply chain management practices and institutional pressures. Sustainability Accounting. *Management and Policy Journal*, *11*(2), 451–476. DOI: 10.1108/SAMPJ-02-2019-0047

Khan, S. A., & Qianli, D. (2017). Impact of green supply chain management practices on firms' performance: An empirical study from the perspective of Pakistan. *Environmental Science and Pollution Research International*, *24*(20), 16829–16844. DOI: 10.1007/s11356-017-9172-5 PMID: 28573559

Khor, K. S., Thurasamy, R., Ahmad, N. H., Halim, H. A., & May-Chiun, L. (2015). Bridging the gap of green IT/IS and sustainable consumption. *Global Business Review*, *16*(4), 571–593. DOI: 10.1177/0972150915581101

Lei, C. F., Ngai, E. W., Lo, C. W., & See-To, E. W. (2023). Green IT/IS adoption and environmental performance: The synergistic roles of IT–business strategic alignment and environmental motivation. *Information & Management*, *60*(8), 103886. DOI: 10.1016/j.im.2023.103886

Lei, J. (2024). Efficient Strategies on Supply Chain Network Optimization for Industrial Carbon Emission Reduction. *arXiv preprint arXiv:2404.16863*. Disponible en: https://arxiv.org/abs/2404.16863

Lembcke, T. B., Herrenkind, B., Nastjuk, I., & Brendel, A. B. (2021). Promoting Business Trip Ridesharing with Green Information Systems: A Blended Environment Perspective. *Transportation Research Part D, Transport and Environment*, *94*, 102795. DOI: 10.1016/j.trd.2021.102795

Li, S., Jayaraman, V., Paulraj, A., & Shang, K. (2016). Proactive environmental strategies and performance: Role of green supply chain processes and green product design in the Chinese high-tech industry. *International Journal of Production Research*, *54*(7), 2136–2151. DOI: 10.1080/00207543.2015.1111532

Li, Y., Pinto, M. C. B., & Diabat, A. (2020). Analyzing the critical success factor of CSR for the Chinese textile industry. *Journal of Cleaner Production*, *260*, 120878. DOI: 10.1016/j.jclepro.2020.120878

Lo, F. Y., & Liao, P. C. (2121). Rethinking financial performance and corporate sustainability: Perspectives on resources and strategies. Technological Forecasting and Social Change, 162, 120346. https://doi.org/DOI: 10.1016/j.techfore.2020.120346

Loeser, F., Recker, J., vom Brocke, J., Molla, A., & Zarnekow, R. (2017). How IT executives create organizational benefits by translating environmental strategies into Green IS initiatives. *Information Systems Journal*, *27*(4), 503–553. DOI: 10.1111/isj.12136

Löser, F. (2015). Strategic Information Systems Management for Environmental Sustainability: Enhancing Firm Competitiveness with Green IS. Vol (6), Universitätsverlag der TU Berlin. https://doi.org/http://dx.doi.org/10.14279/depositonce-4447

Mahraz, M. I., Benabbou, L., & Berrado, A. (2020). A Compilation and Analysis of Critical Success Factors for the ERP Implementation. *International Journal of Enterprise Information Systems*, *16*(2), 107–133. DOI: 10.4018/IJEIS.2020040107

Martínez-Arroyo, J. A., Valenzo-Jiménez, M. A., & Kido-Cruz, A. (2022). Gestión ambiental, innovación y ventaja competitiva en empresas exportadoras de aguacate. *Investigación administrativa*, 51(129). https://doi.org/DOI: 10.35426/IAv51n129.01

Masa'deh, R., Alananzeh, O., Algiatheen, N., Ryati, R., Albayyari, R., & Tarhini, A. (2017). The impact of employee's perception of implementing green supply chain management on hotel's economic and operational performance. *Journal of Hospitality and Tourism Technology*, *8*(3), 395–416. Advance online publication. DOI: 10.1108/JHTT-02-2017-0011

Meacham, J., Toms, L., Green, K.Jr, & Bhadauria, V. (2013). Impact of information sharing and green information systems. *Management Research Review*, *36*(5), 478–494. DOI: 10.1108/01409171311327244

Melville, N. (2010). Information Systems Innovation for Environmental Sustainability. *Management Information Systems Quarterly*, *34*(1), 1. DOI: 10.2307/20721412

Molla, A., Cooper, V., & Pittayachawan, S. (2011). The Green IT Readiness (G-Readiness) of Organizations: An Exploratory Analysis of a Construct and Instrument. *Communications of the Association for Information Systems*, *29*(4), 67–96. https://aisel.aisnet.org/cais/vol29/iss1/4. DOI: 10.17705/1CAIS.02904

Nanath, K., & Pillai, R. R. (2017). The Influence of Green IS Practices on Competitive Advantage: Mediation Role of Green Innovation Performance. *Information Systems Management*, *34*(1), 3–19. DOI: 10.1080/10580530.2017.1254436

Norheim, H. (2018). The virtues of green strategies: Some empirical support from the Alliance context. *Journal of Business Ethics, 151*(4), 1161–1173. DOI: 10.1007/s10551-016-3274-8

Pérez, L. M. (2021). *Factores críticos de éxito y fracaso de la transformación digital en las empresas. Tesis de maestría, Universidad de Talca.*https://dspace.utalca.cl/bitstream/1950/13204/3/2021A001044.pdf

Petter, S., DeLone, W. H., & McLean, E. R. (2013). Information Systems Success: The Quest for the Independent Variables. *Journal of Management Information Systems, 29*(4), 7–62. DOI: 10.2753/MIS0742-1222290401

Pooja, P. (2024). Environmental Degradation: Causes, Impacts and Mitigation. *Journal of Advances in Science and Technology, 20*(1), 220–223. DOI: 10.29070/94vmdz61

Purnomo, A., Susanti, M., & Widodo, W. (2022). *A study of green information technology using bibliometric analysis.* Springer Nature Singapore Pte Ltd., DOI: 10.1007/978-981-16-9573-5_46

Qu, T., & Liu, Y. (2022). Green innovations, supply chain integration and green information system: A model of moderation. *Journal of Cleaner Production, 340*, 130667. DOI: 10.1016/j.jclepro.2022.130557

Quiroz, D., & Vera, C. (2023). *Ética ambiental corporativa e innovación verde: Un modelo de gestión empresarial. Revista Facultad de Ciencias Económicas, 31*(1), 75-92. https://revistas.unimilitar.edu.co/index.php/rfce/article/view/6715/5862

Rajan, C. A., & Baral, R. (2015). Adoption of ERP system: An empirical study of factors influencing the usage of ERP and its impact on end user. *IIMB Management Review, 27*(2), 105–117. DOI: 10.1016/j.iimb.2015.04.008

Ram, J., & Corkindale, D. (2014). How "critical" are the critical success factors (CSFs)? Examining the role of CSFs for ERP. *Business Process Management Journal, 20*(1), 151–174. DOI: 10.1108/BPMJ-11-2012-0127

Reitsma, E., & Hilletofth, P. (2018). Critical success factors for ERP system implementation: A user perspective. *European Business Review, 30*(3), 285–310. DOI: 10.1108/EBR-04-2017-0075

Ruiz, E. I. M., Almazán, D. A., & Cortés, C. B. Y. (2023). Multidimensional scale for assessing the determinants of success of green information systems on organizational performance. *Revista del Centro de Investigación de la Universidad la Salle, 15*(59), 105–130. DOI: 10.26457/recein.v15i59.3524

Ryoo, S. Y., & Koo, C. (2013). Green practices-IS alignment and environmental performance: The mediating effects of coordination. *Information Systems Frontiers*, *15*(5), 799–814. DOI: 10.1007/s10796-013-9422-0

Saade, R. G., & Nijher, H. (2016). Critical success factors in enterprise resource planning implementation: A review of case studies. *Journal of Enterprise Information Management*, *29*(1), 72–96. DOI: 10.1108/JEIM-03-2014-0028

Sarkar, P., & Young, L. (2009). Managerial attitudes towards Green IT: An explorative study of policy drivers. PACIS 2009 Proceedings, 1–14. https://aisel.aisnet.org/cgi/viewcontent.cgi?article=1071&context=pacis2009

Saygili, E. E., Ozturkoglu, Y., & Kocakulah, M. C. (2017). End users' perceptions of critical success factors in ERP applications. *International Journal of Enterprise Information Systems*, *13*(4), 58–75. DOI: 10.4018/IJEIS.2017100104

Seidel, S., Bharati, P., Watson, R. T., & Boudreau, M. (2017). The Sustainability Imperative in Information Systems Research. *Communications of the Association for Information Systems*, *40*(1), 40–52. Advance online publication. DOI: 10.17705/1CAIS.04003

Suša Vugec, D., Bosilj Vukšić, V., Pejić Bach, M., Jaklič, J., & Indihar Štemberger, M. (2020). Business intelligence and organizational performance: The role of alignment with business process management. *Business Process Management Journal*, *26*(6), 1709–1730. DOI: 10.1108/BPMJ-08-2019-0342

Talib, M. S., Hamid, A. B. A., & Thoo, A. C. (2015). Critical success factors of supply chain management: A literature survey and Pareto analysis. *EuroMed Journal of Business*, *10*(2), 234–263. DOI: 10.1108/EMJB-09-2014-0028

Velásquez, T., Messino, A., Navarro, H. E., & Ovalle, A. A. (2019). The role of Green IT in the non-functional software development requirements: Perspectives for functional design. *Journal of Physics: Conference Series*, *1388*(1), 012030. DOI: 10.1088/1742-6596/1388/1/012030

Watson, R. T., Boudreau, M., & Chen, A. J. (2010). Information Systems and Environmentally Sustainable Development: Energy Informatics and New Directions for the IS Community. *Management Information Systems Quarterly*, *34*(1), 23. DOI: 10.2307/20721413

Wu, S., & Wu, Y. (2014). The influence of enterprisers' green management awareness on green management strategy and organizational performance. *International Journal of Quality & Reliability Management*, *31*(4), 455–476. DOI: 10.1108/IJQRM-01-2013-0019

Yang, Z., Sun, J., Zhang, Y., & Wang, Y. (2017). Green, Green, It's Green: A Triad Model of Technology, Culture, and Innovation for Corporate Sustainability. *Sustainability (Basel)*, *9*(8), 1369. DOI: 10.3390/su9081369

Yang, Z., Sun, J., Zhang, Y., & Wang, Y. (2018). Peas and carrots just because they are green? Operational fit between green supply chain management and green information system. *Information Systems Frontiers*, *20*(3), 627–645. DOI: 10.1007/s10796-016-9698-y

Yang, Z., Sun, J., Zhang, Y., & Wang, Y. (2019). Perceived fit between green IS and green SCM: Does it matter? *Information & Management*, *56*(7), 103154. Advance online publication. DOI: 10.1016/j.im.2019.02.009

Yohannes, A., & Mauritsius, T. (2022). Critical Success Factors in Information Technology Projects. *International Journal of Emerging Technology and Advanced Engineering*, *12*(7), 45–67. DOI: 10.46338/ijetae0722_06

Zhang, Y., & Wei, F. (2021). SMEs' charismatic leadership, product life cycle, environmental performance, and financial performance: A mediated moderation model. *Journal of Cleaner Production*, *306*, 127147. DOI: 10.1016/j.jclepro.2021.127147

Chapter 7
Exploring the Convergence of Artificial Intelligence and Sustainable Computing

C. Geetha
https://orcid.org/0000-0002-9894-1524
R.M.K. Engineering College, India

T. Chandrasekar
https://orcid.org/0000-0003-3591-2205
Kalasalingam Academy of Research and Education, India

K. Sujatha
https://orcid.org/0000-0002-7648-342X
SRM Institute of Science and Technology, Ramapuram, India

S. D. Lalitha
https://orcid.org/0000-0003-3712-8254
R.M.K. Engineering College, India

A. Siva Kumar
https://orcid.org/0000-0001-8205-1407
SRM Institute of Science and Technology, Kattankulathur, India

S. Nithi Ravya
https://orcid.org/0009-0002-9433-8258
Sri Sairam Engineering College, India

ABSTRACT

Achieving sustainability and reducing inequities are at the heart of the 17 Sustainable Development Goals set out by the United Nations. When

DOI: 10.4018/979-8-3693-8643-9.ch007

faced with complicated problems, a multidisciplinary team using data-driven methodologies, artificial intelligence, and technology may maximise efficiency, incorporate sustainability considerations, and back up well-informed decisions. This study classifies AI-driven solutions to sustainability problems by doing a literature review, scientometric analysis, and semantic analysis. It emphasises the interface of artificial intelligence and sustainability, as well as important research subjects and regional variations. Based on the results, it's clear that multi-dimensional decision-making requires hybrid methods that incorporate AI, data analytics, and human expertise. Also, the report stresses the significance of making sure AI and big data are used sustainably and the ethical considerations that come with it. To fill in knowledge gaps and propel successful sustainability solutions, collaboration and inclusive research are crucial.

1. INTRODUCTION

In the years since the United Nations (UN) adopted its 17 Sustainable Development Goals (SDGs), the issue of sustainability has grown in importance. These objectives provide a world-wide road map for fixing our most serious economic, social, and environmental problems (Bexell M et. al, 2017; Biermann F, 2017). The time for action is now, more than ever, as communities throughout the globe work towards sustainable development and as shown in Figure 1. The SDGs stress the importance of coordinated international and regional actions to address inequality and its wider social and environmental impacts (Coulson RN et. al, 1989; Haefner N et. al 2021). Encouraging policies and frameworks that integrate sustainability principles across all sectors, they push for proactive actions to achieve a fair and equitable world (Bachmann N et. al, 2023).

Figure 1. Sustainable Development Goals

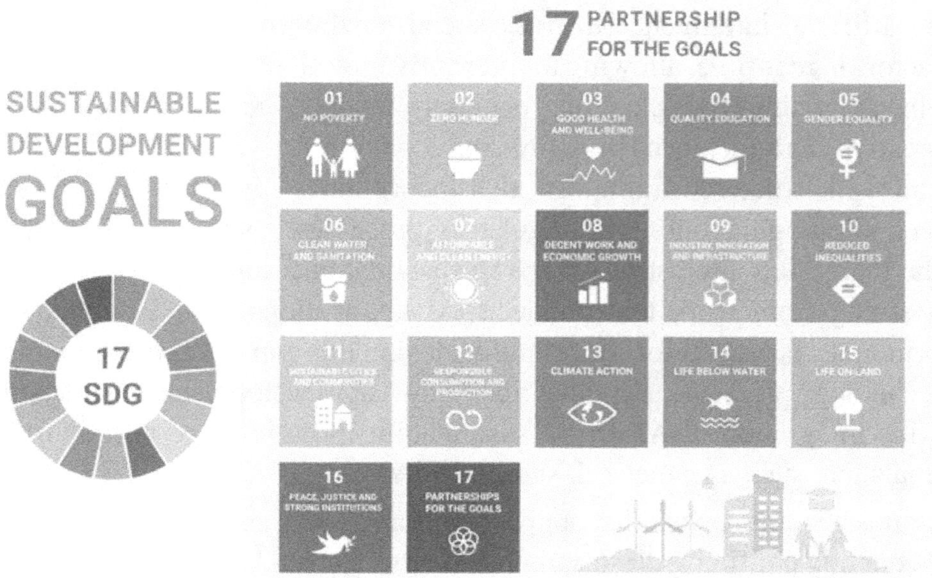

Courtesy: https://www.energeia.in/walking-the-talk-with-sustainable-development-goals/

To address sustainability-related challenges, recent technological advances have become crucial facilitators. A growing number of experts are beginning to acknowledge the possibilities presented by data-driven approaches, the Internet of Things (IoT), ICT, blockchain, sensors, Artificial Intelligence (AI), and Big Data (BD). By improving stakeholder engagement, streamlining resource management, speeding up innovation, encouraging cross-industry applications, and allowing for long-term sustainable planning, these technologies make it easier to integrate different aspects of sustainability (Kaack LH et. al, 2022). Artificial intelligence (AI) and data-driven solutions can lessen environmental degradation and social and economic inequality through a methodical and collaborative strategy (Dhar P, 2020).

Sustainability projects can be made possible through the integration of AI and data-driven techniques (Salvia AL et. al, 2019) by 1) Creating smarter ways of consuming energy and resources that are both efficient and environmentally friendly. 2) Putting into place policies backed by evidence that make use of insights generated by AI to improve governance, regulatory frameworks, and decision-making.3) Using analytics powered by artificial intelligence to track social, environmental, and economic factors in real-time, allowing for preventive solutions. 4) Reducing carbon emissions, improving circular economy models, and minimising waste creation can be achieved by optimising economic activities.5) A systematic and well-structured framework that incorporates sustainability policies, technology guidelines, and regulatory processes is still lacking, despite the increasing interest in AI-driven sustainability solutions. 6) Building a strong framework that connects AI applications with sustainability principles is crucial for successfully closing this gap. 7)A carbon-neutral technology promotion, Circular Economy (CE) development, and industry-wide integration of AI-driven sustainability solutions are all important goals that such a framework should address.

It is becoming clear that our present frameworks and techniques, which frequently conflict with sustainable ideals, may not be able to completely realise the solutions we need in our pursuit of sustainable AI (Heilinger, J.-C et. al, 2023). It should be borne in mind that sustainability is still an unrealised goal, a vision held by many governments and society (Kemp et. al, 2007; Mensah. J et. al, 2019). It should come as no surprise, then, that there is no agreed-upon definition. This journey calls on us to venture out of our comfort zones and discover "other worlds" in our pursuit of this goal. The natural environment, with its built-in systems of balance and renewal, other traditions' time-tested philosophies, or even a future scientific vision that questions our current standards could all fit this description (Kemp and Martens (2007) describe sustainability science as a new kind of science). The only way to find out about these "other worlds" that may have solutions to our environmental problems is to broaden our minds. If we do this, we may be able to find new ways of thinking and information that could help us build AI systems that can last.

Artificial intelligence (AI) is a hot topic in the business world. Healthcare, transportation, agriculture, hospitality, manufacturing, and education are just a few of the many industries that are feeling the increasing influence of AI. As someone who uses ChatGPT or has noticed AI's expanding influence in their personal and professional lives, you have likely seen the rising significance of AI directly.

In recent years, the AI market has experienced tremendous growth. Statista estimates that the artificial intelligence business was worth more than $184 billion in 2024, up about $50 billion from the previous year. By 2030, experts predict that the sector would have grown at a staggering rate, topping 826 billion USD.

Figure 2. Global AI Market Growth Forecast (2020-2030)

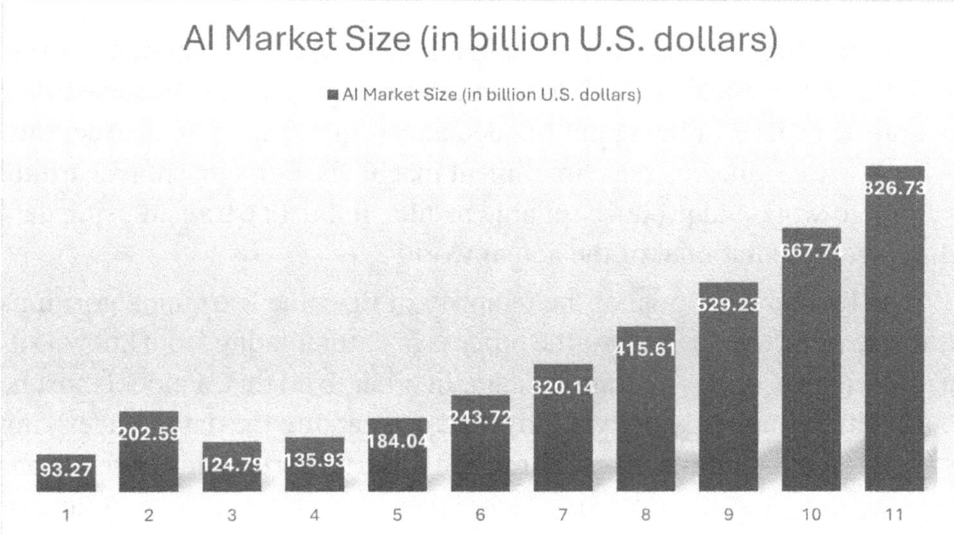

From an estimated $93.27 billion in 2020 to $826.73 billion by 2030, the data shown in Figure 2 implies that the Artificial Intelligence (AI) market is growing at an exponential rate. The escalating integration of AI across several areas, such as healthcare, banking, manufacturing, and education, is evident in this rapid expansion.

Climate monitoring, smart resource management, and energy efficiency are three areas where AI could lead to major improvements from a sustainability perspective. Industries can benefit from AI-powered solutions like optimisation algorithms, predictive analytics, and supply chain efficiency boosts, all of which contribute to a decrease in carbon footprints. High energy usage, technological waste, and ethical AI practices are some of the concerns that make AI sustainability a challenge in and of itself.

A sustainable future dependent on AI's continued development necessitates a balance between economic advantages and environmental responsibilities.

2. THE INVESTMENT IN TIME AND EXPENSE NECESSARY TO TRAIN AI MODELS

A key component of artificial intelligence (AI) is machine learning, which uses statistical methods to train computer programs to become better over time instead of being hard-coded according to specific instructions. AI as a whole aims to create intelligent machines. For a machine learning system to work and produce reliable results, it must be trained using data that is representational of the actual world.

The fundamental goal of the majority of machine learning algorithms is to use inductive learning—the process of extrapolating from known situations to new ones—to training data. In order to do this, a model must be built, either implicitly or explicitly, that can capture the data's underlying structures and patterns. Algorithms like this derive rules and correlations from newly input data by using techniques like clustering, regression, or classification. This allows them to make judgements or predictions based on the new data. Importantly, a good model's capacity to generalise means that it has learnt more than only the training data (fitting), but rather a simplified version of the fundamental tendencies in the data that can be applied to the wider environment it represents. Inductive learning is a cornerstone of machine learning's predictive abilities since it allows the model to generalise its findings from a sample to the full population.

A subset of machine learning, deep learning is defined by its ability to sift through and draw conclusions from massive datasets by means of multi-layered neural networks. In contrast to more conventional ML methods, deep learning architectures can find features on their own. These networks mimic the decision-making process by simulating it using layers of interconnected nodes, also called neurones. Data moves across these layers, with deeper layers capturing more abstract and complicated representations and each layer extracting higher-level attributes. In applications like picture and speech recognition, where the intricacies of patterns are crucial, deep learning models perform exceedingly well because of this hierarchical learning strategy. Deep learning models are also very good at dealing with high-dimensional and non-linear data, which makes them a powerful tool for solving many complicated issues that computers couldn't handle before. The remarkable pattern recognition capabilities of deep learning models are propelling developments across numerous domains, and they are quickly becoming a cornerstone of artificial intelligence.

The extraordinary complexity, flexibility, and potential for autonomy of AI set it distinct from other technologies. Artificial intelligence (AI) stands out from other technologies because of its capacity to learn from data, make judgements, and execute activities that have historically required human knowledge. However, this capability does not alleviate all environmental and societal challenges. Artificial intelligence systems have progressed to the point that they can not only carry out predetermined procedures, but also improve their performance over time via learning, eventually outperforming humans in some areas.

Beyond the scope of specific jobs, AI is having systemic effects. Its ability to optimise whole networks, including energy grids and transportation systems, with substantial efficiency benefits is a result of its integrative nature. There is a higher carbon footprint associated with AI because of these optimisations, but there is also a caveat: greater energy demands for processing enormous volumes of data and potential unexpected consequences stemming from complex AI behaviours (Brevini et. al, 2021). Furthermore, AI has the potential to unintentionally reinforce biases revealed in training data, which might cause ethical quandaries and governance issues that have never been seen with prior technologies (Sison et. al, 2023). Both exciting and troubling new complexities have emerged because of applying machine learning to sustainability issues. Firstly, these algorithms may be able to solve the complex, non-linear, multi-variable issues that crop up in sustainability-related fields as ecological conservation, supply chain management, and resource allocation (Dilkina, B. et. al, 2017; Fisher, D et. al, 2017). On the other hand, there are difficulties associated with these algorithms' most strong properties, which are their capacity to take into account numerous dimensions at once.

There is some evidence that larger machine learning models (those with more parameters and layers) provide better results. Nevertheless, there is a growing monetary and ecological price to pay for this advancement (Brown R. E, et. al, 2007). Machine learning models tend to grow at a faster rate than their performance improves, suggesting that the high costs of doing so will likely remain. There are growing concerns about the amount of energy computers use, which means that energy expenses could end up being the most important part of the overall cost of ownership (Brown, R.E et. al, 2007). Capital expenditures, operational outlays, environmental repercussions, and energy management are all factors to be considered when running a data centre (Barroso, L.A. et. al, 2007).

Factors such as the size and complexity of the dataset, the optimisation algorithm utilised, and the model's design all contribute to the computing cost of training a machine learning model. A loss function is a mathematical formulation used in artificial neural networks (ANNs), a class of machine learning algorithms (shown in Figure 3), to quantify the discordance between the network's forecasts and the true target values in the training data. The loss function measures the degree to which the model closes the gap between the inputs and outputs of the underlying function. By manipulating the weights of the network, the purpose of optimisation is to minimise the loss function. This process of fine-tuning requires a lot of calculations, which means it takes a lot of time to process (Garbin et. al, 2020).

Figure 3. Deep Learning as it pertains to AI

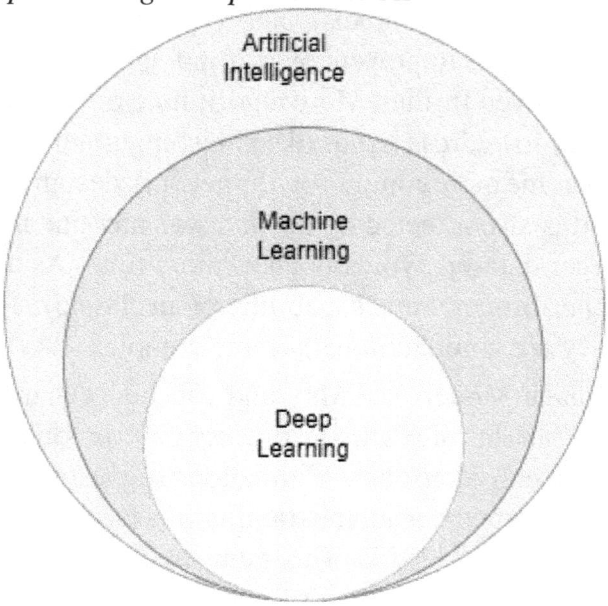

When it came to classic machine learning models, the size of the training set was the main factor that dictated complexity (Ofek N et. al, 2017). Now, however, architectural intricacies are being hailed as a key factor in computational complexity, thanks to the proliferation of Deep Neural Networks (DNNs), a subset of ANN specifically engineered to model and comprehend intricate patterns and representations with multiple input and output layers. DNNs are enhanced in representational power but pose computational issues due to their various hidden layers, which sometimes include specialised types such recurrent layers and convolutional (Ofek et. al, 2017).

Stochastic Gradient Descent (SGD) and other optimisation methods used to train DNNs traverse a much broader parameter space as they iteratively update millions or even billions of parameters while minimising the loss function. Because these algorithms typically need more iterations to reach a perfect solution, the computing effort increases as a consequence. In comparison to more conventional models, each iteration requires the computation and updating of a much bigger set of parameters. As the number of neurones and their connections increases in DNNs, the training phase forward and backward passes become more resource intensive (Zhou et. al, 2021). In addition, complex DNNs typically call for sophisticated regularisation strategies to prevent overfitting, which increases the computational burden even further. More algorithmic or architectural tweaks may be necessary to address issues like vanishing gradients (Hanin B. et. al, 2018) that become more common as the network design gets deeper. The computing load is also affected by the dataset's amount and complexity; training on larger datasets typically takes more time. As a result, DNNs provide better performance and capabilities than classic machine learning models, but they are computationally more complex.

Large Language Models (LLMs), and especially neural transformer models, are using an alarming amount of energy and producing an alarming amount of carbon emissions. New estimations suggest that 280,000 kg of CO2 could be consumed during the training of a large neural transformer model (Yang, T.-J. et. al, 2017). This number is worrisome both in isolation and when put in the context of people's individual and collective scientific efforts, even though it is difficult to evaluate the emissions of these models (Patterson, D et. al, 2021). As an example, it is commonly mentioned in talks about personal and business carbon footprints that a transatlantic flight from London to New York releases roughly 900 kg of CO2 per passenger. Thus, more than 300 transatlantic flying passengers' emissions are equal to training one huge neural network. So, many NLP projects are responsible for a significant amount of greenhouse gas emissions.

The View of Sustainability via the Lens of Complexity

Most machine learning models have computational complexity that grows exponentially with the number of features or samples, which is an important fact to keep in mind. Particularly when working with massive datasets, this increasing complexity is a major cause for alarm. Dimensionality partitioning in a mutually exclusive manner is one effective approach to mitigate this issue. Splitting the issue space into smaller, mutually exclusive subsets is a great way to lessen the computing load.

One way to break the problem down into smaller, more manageable pieces is to use a divide-and-conquer algorithm (Hsieh, C.-J et. al, 2014) or an ensemble method (Ganaie, M.A et. al, 2022). These smaller problems can be handled separately, maybe even simultaneously, and then integrated to create the larger answer. The algorithm's computing complexity is drastically reduced, making it more efficient and scalable, thanks to this methodological adjustment, which also distributes the computational strain. Thus, these partitioning methods provide a potential option to handle learning algorithms that are computationally costly, helping with scalability and making good use of computational resources. There is an urgent and complicated need to address the issue of scalability when applying machine learning algorithms to massive datasets. One the one hand, businesses in all sorts of industries are amassing mountains of data, from consumer habits to scientific findings, that could be useful for drawing conclusions. Contrarily, not all ML algorithms are well-suited to handle datasets of this size. There are a lot of reasons to fix this scalability problem.

A prominent explanation is that the learnt models' accuracy is directly proportional to the size of the training set. In situations when the feature space is huge or where the software needs to learn "small disjuncts" or unusual examples that are required for high accuracy, overfitting can occur with smaller datasets (Provost, F et. al, 1999). When data is noisy, it becomes even more difficult to tell real outliers from real special situations. The computational complexity of learning algorithms is another aspect of the scalability challenge. Increases in dataset size can render algorithms with greater-than-linear complexity unworkable. This isn't only a problem with predictive modelling; it affects data-mining applications that aim to unearth new information as well. Nevertheless, in order to accomplish this without being swamped by false minor disjuncts, datasets big enough to allow for confident generalisation are required. So, it's crucial to find ways to make algorithms that can process massive information quickly and scalable without sacrificing accuracy or making them harder to understand (Nguyen, A.-P. et. al, 2023)

Many data representation approaches, particularly those dealing with text mining, make use of sparse feature representations (Ofek, N. et. al, 2020). Each feature, for instance, is associated with the frequency of a particular n-gram in the conventional term frequency (TF) based approach to text representation. Traditional text classification tasks, such as sentiment analysis, have shown great success with this simple method (Cao, T. et. al, 2015). The "curse of dimensionality" and other problems may arise with this approach's high-dimensional feature space. The computing cost of finding the optimal solution grows exponentially with the number of dimensions, which are variables or restrictions. This makes the model susceptible to overfitting and creates situations where optimisation tasks become computationally impossible, reducing the algorithms' usefulness in addressing large-scale sustainability concerns in the actual world (Bengio, Y. et. al, 2005).

3. CONCERNS FOR ETHICS AND THE ENVIRONMENT IN THE CONTEXT OF SUSTAINABLE AI

In this section, we will examine AI systems that are specifically designed to tackle sustainability-related problems, expanding on our earlier discussion of the time, money, and resources needed to construct any AI system. A troubling irony is highlighted by this paradox: the artificial intelligence (AI) technologies that are being used to accomplish the SDGs and safeguard our ecosystems could unintentionally contribute to their degradation (Nordgren, A. et. al, 2023; Cowls, J. et. al, 2023). As a result, we label technologies as sustainable even though there is no need to do so.

The Environmental and Moral Expenses

Ecological system monitoring, intervention impact simulation, and sustainable resource allocation are all areas where AI models have demonstrated potential. It is intriguing that marginalised populations are frequently the ones that do not reap the benefits of technology developments, even if deploying large AI models comes with substantial computational and environmental costs (Bender, E.M. et. al, 2021). The brunt of climate-related dangers is frequently felt by these populations (Bullard, R.D. et. al, 1993). Multiple research confirm that low-income households often live in places that are more susceptible to many climate-related hazards, such as flooding, water contamination, excessive heat, and mudslides. Because of their low adaptive ability, these communities are already at a heightened risk of experiencing even more severe social inequality as a result of these conditions. Keep in mind that SDG 10 (reduced inequities) and the other SDGs aim to do just that—reduce inequality.

According to reports, 11% of the world's population lived in low-elevation coastal zones in 2000. Many of these people were economically disadvantaged and restricted to flood-prone locations, further illustrating the impact of climate change on marginal communities (un.org report on 2023). The South and East Asian, Latin American, and Caribbean regions exhibit this tendency to a comparatively greater degree. Climate change is making life even more difficult for the approximately 29 percent of the world's population who resides in dry sub-humid zones, semi-arid zones, or completely dry areas. The fact that different socioeconomic groups are vulnerable to different degrees adds complexity to these structural inequities. The impoverished in Mumbai, India, for instance, fork out a larger share of their income fixing up their homes after floods than the well-off. Just as Hurricane Katrina in 2005 devastated New Orleans, it also highlighted how low-income African Americans were more vulnerable due to intersecting disparities characterised by race, education, and income, which made recovery much more difficult for this demographic.

A feeling of extreme urgency is being fostered in the global community by the United Nations Agenda 2030, which specifically requests "urgent action to combat climate change and its impacts" and "...transformative steps that are urgently needed to shift the world onto a sustainable and resilient path" (Cui, S et. al, 2023). In light of this growing sense of urgency, artificial intelligence (AI) is being increasingly considered as a potential solution to our complex social and environmental problems. The increasing urgency to accomplish these objectives within the allotted time may lead some to view AI as a "magic wand" that can magically fix problems with complicated systems (Berti-Equille, L. et. al, 2021). Despite the fact that AI presents exciting opportunities to speed up the transition to sustainability, this view runs the risk of oversimplifying the complex ethical and practical issues that come with implementing AI solutions on a massive scale. Thus, it is crucial to exercise caution so that the cautious and ethical use of AI in the quest for global sustainability is not jeopardised by this feeling of urgency.

The human desire to manipulate the natural world, which stems from primal urges and is exacerbated by contemporary arrogance, sheds more light on this conundrum. This endeavour is clearly visible in the field of artificial intelligence (AI) in the numerous attempts to control human behaviour and social dynamics (Portier, K. et. al, 2013) and to manage the complexities of natural ecosystems (Wu, X et. al, 2018). Unintentionally, AI has the potential to fuel nationalism, amplify biases, and violate human rights, especially in places where democratic supervision, ethical scrutiny, and transparency are lacking (Vinuesa, R et. al, 2020). Meanwhile, the CalGang database has been criticised for its disproportionate bias and pervasive inaccuracies (Crawford, K. et. al, 2021), and the development of "citizen scores" (Nagler, J. et. al, 2019)aims to control social conduct through AI-driven analytics.

In light of the Anthropocene, a word that describes the enormous influence that humans have had on Earth, the idea of gaining more control over both natural and human systems by technological and scientific interventions is controversial (Coeckelbergh, M. et. al, 2021). This begs the critical question: does increasing human control over Earth's processes line up with the goal of harmony, or does it make things worse? To determine if the idea of enhancing human agency in relation to ecosystems and society is a continuation of the problems associated with the Anthropocene epoch or a move towards remediation, it requires careful examination.

Sustainable AI Metrics

There is an increasing need for new evaluation criteria to be created and used (Schwartz, R. et. al, 2020). A model's performance, computational efficiency, and environmental effect are only a few of the many elements that these metrics should represent. To help researchers and policymakers make better, more informed decisions about whether and how to invest in new model training or upgrades, we may combine these elements into a uniform evaluation framework.

Evaluating other parameters, such as model size (storage cost) and computational operations (throughput), is easier than gauging the energy expense associated with DNNs.

Data transfer accounts for a substantial portion of energy consumption, which is difficult to measure due to factors outside the control of the DNN's design. This adds another layer of complexity to the problem. In their approach to estimating the DNN's energy usage, Yang and colleagues (2017) take into account the structural architecture of the network as one of several criteria.

The Pareto frontier can be used to define the optimal trade-off between cost and prediction performance, delineating a boundary in the objective space where any further improvement in one metric would necessitate a compromise in the other. This frontier serves as a valuable guide for decision-makers since it allows them to balance the computational cost of an algorithm against its predictive accuracy, facilitating more informed and sustainable choices in model selection and deployment. For example, Ofek and colleagues (2018) examine points along the Pareto frontier to identify algorithms that offer the most favorable trade-offs, thereby optimizing the allocation of computational resources without sacrificing the quality of predictions.

4. PHILOSOPHICAL AND ECOLOGICAL FRAMEWORKS FOR SUSTAINABILITY

Finally, the part provides a critical analysis of the problems with a sustainability checklist method (Bolte, L. et. al, 2022), which overlooks the fundamental goals of sustainability and brings up questions of fairness. Given the complexity and interdependence of our world, which cannot be adequately addressed by accomplishing a singular objective, this omission should not come as a surprise.

Given the need for a more comprehensive set of principles that represent a unified vision in this work, we will now examine the frameworks that can improve the creation of sustainable AI, and then we will discuss how to put these ideas into practice when dealing with AI. Adaptability, balance, harmony, and a recognition of interconnectedness are qualities inherent to sustainability, and they must be incorporated into the design of AI systems. This is in line with our stance, which differs from the dominant narrative of tension and argues that harmony is the foundation of sustainable AI (Van Wynsberghe, A. et. al, 2021).

To what extent do the aforementioned frameworks constitute this type?

What follows is an ecological and complex systems theory–based philosophical framework. Aspiring for harmony and balance, these frameworks offer a different way of seeing and engaging with the world, which is why they are considered alternatives.

The Ecological Principle: A Path to Ethical Sustainability

While the ethics of sustainability is still up for debate, this section focusses on outlining a path towards ethical sustainability by placing an emphasis on the ecological principle. The fundamental and foundational value of ecosystems, as well as their contribution to human prosperity, are argued to outlive their physical presence in certain works that propose sustainability as an essential ethical principle. It acknowledges that we may change unsustainable practices by making a conscious effort (Pierce, J et. al, 2004). Because sustainability is multidimensional, another argument is that complexity is a necessary condition for ethical sustainability (Asheim, G.B et. al, 1994).

We focus on the ecological principle as the core of ethical sustainability instead of trying to define it strictly. In accordance with this idea, we must have a holistic view that highlights the interdependence of all living beings, particularly those at the bottom of the social ladder. In light of this diversity of life and the surroundings in which it exists, including human beings and the social constructions around them, the ecological principle stresses the need of harmony and balance. In accordance with this idea, ethical considerations must take into account the possibility of social inequities, economic ramifications, and the lives of non-human entities while assessing the effects of progress. Ethical sustainability necessitates a comprehensive understanding of these complex interrelationships; optimising isolated factors, like carbon emissions, won't cut it; rather, we need to strike a balance that fosters equitable and long-term solutions.

Because it broadens the ethical vision to incorporate multiple viewpoints, this ecological principle should play a central role in sustainability engineering and science. It addresses potential biases against minorities that AI systems may perpetuate and integrates issues regarding privacy, autonomy, as well as regulate (Sison, A.J.G. et. al, 2023). By making sustainability efforts centred around the ecological principle, we can make sure that the path of AI development and deployment is in line with both economic and ethical considerations for the larger community it is supposed to assist (Nash, K.L. et. al, 2020).

The Paradigm of Self-Organization

An important aspect of sustainability discourse is the incorporation of complex systems, which highlights the complex interaction between agents that can exchange and adjust information dynamically (Bolte, L et. al, 2022). As we explore the inner workings of these systems, we discover that self-organization (Seeley, T.D., 2002) is fundamental to complex systems and provides a strong contrast to the current tendency to manipulate natural processes through technology. Ecological models that stress the organic development of order in the absence of strict regulation are consistent with this procedure, which is based on the ideas of decentralisation and emergent behaviour. Because the harmonious essence of sustainability is precisely this giving up of control, self-organization provides a suitable platform for expressing our hopes for a sustainable future.

Anthropocene assumptions about human dominion and the idea that artificially created algorithms should replace natural systems' self-regulating mechanisms are rebuked by self-organization. The interventionist strategy of algorithmic ecosystem optimisation is called into question by this paradigm because it can fail to take into consideration the all-encompassing wisdom inherent in natural processes. Our relationship with artificial intelligence (AI) is being rethought through the promotion of decentralised models by self-organization. Rather than imposing strict control, we are now encouraged to work together to coexist with the intricacies of social and natural systems. Sustainability, according to this view, is best achieved by working in harmony with, rather than trying to control, the natural order of things. This method recognises the complex governance that exists in nature.

In particular, "self-organization" describes the newly developed structures that result from networks of agents (like beehives) working together and coordinating their activities in the absence of a commanding queen-bee or other centralised manager. Because they draw on the expertise and insights of a wide range of people, these self-organising systems are naturally better at making decisions and other complex tasks. Unlike centrally managed systems (such lawmakers employing AI models for decision support) or isolated autonomous systems (where each entity functions autonomously), self-organising networks take advantage of decentralised behaviour. These networks can modify their behaviour to meet common demands and environmental differences in a sustainable and efficient way. Research shows that these decentralised systems can perform better than their more traditional hierarchical equivalents in some situations. For example, research indicates that autonomous systems that are either geographically dispersed or centrally controlled aren't as good at water management as farmer networks that are self-organising (Lansing, J.S. et. al, 1993). This perspective runs counter to the common belief that optimising resource management is always better than using technology interventions, which are frequently driven by AI.

Simple Revolution

An illuminating prism through which to view sustainability concepts is the Daoist tenet of harmony, balance, and intentional non-action (Nelson, E.S. et. al, 2020). The ancient Chinese philosophy of Daoism stresses the unity of all things and calls for humans to live in harmony with nature. This philosophy provides an alternate framework that expands upon current sustainability paradigms, and it is strikingly congruent with the current demands of sustainable development ((Luo, Q. et. al, 2009). By encouraging actions that are inherently balanced with the rhythms of natural systems, Daoism teaches us to do more than simply optimise; it transforms us. Therefore, Daoism adds depth to the discussion of sustainability by bringing in comprehensive, long-term views that reflect its ancient knowledge, which may cause us to reevaluate our current notion of sustainable living.

A notion that should be defined by the Daoist founder Laozi around 2500 years ago is that human actions should be in harmony with Earth's laws. The idea of "wu-wei," which originally meant "no (wu) effort/work" (wei) in Daoism, is a profound and difficult one. Wu-wei is not intending to sit on his hands. It implies doing less, or more accurately, doing what comes naturally as a result of the situation, without forcing anything (Loy D et. al, 1985). Although the Daoist perspective differs with Agenda 2030's tenets, it does provide an alternate framework that has the potential to significantly transform how sustainability is understood and put into practice (Luo, Q. et. al, 2009). Responsible consumption is the subject of SDG 12 and climate action is the subject of SDG 13. Energy management, optimisation, and waste minimisation have been the traditional foci of these objectives, but the Daoist perspective implies a shift towards a more transformational stance. Goal 12 of the Sustainable Development Agenda would focus on really reducing consumption, rather than just making production and consumption less destructive. This is in line with wu-wei principles: if we do less on purpose, we might be able to accomplish more for the environment and sustainability. Also, the Daoist advice to "act less" could be useful for SDG 13, which demands immediate action to fight climate change. This in no way suggests ignoring climate change; rather, it acknowledges that, rather than rushing in with both feet, sometimes a measured and deliberate approach yields better results. According to Daoist theory, we need to do less of what hurts the earth in order to find climate solutions, rather than more of the same. This includes more technology, more control, and more mitigation.

5. A PARADIGM SHIFT: MOVING BEYOND METRICS

An disconcerting commodification of nature may result from the increasing fixation on metrics for evaluating the impact of AI systems on the environment (Schwartz, R. et. al, 2020). To begin, by reducing the multi-faceted worth of ecosystems and natural resources to purely economic terms, this method makes it easier to acquire and regulate these things through market-based procedures (Acevedo, B. et. al, 2021). Instead of viewing nature as an interdependent system deserving of respect, this reductionist view promotes a transactional paradigm in which nature is seen as a collection of assets to be maximised. Secondly, the complex, adaptive nature of ecological systems poses a significant challenge to the accurate measurement of AI's environmental impact and contribution. Given the intricate web of dependencies and feedback loops that characterize natural ecosystems, it is overly optimistic, if not naive, to presume that all potential ramifications can be captured through quantifiable metrics.

A dependence on metrics not only steers us away from the harmonious approach required for true sustainability, but it also cultivates a narrow view that may overlook the broader, often immeasurable aspects of ecological influence.

In the quest to place AI within the vision of sustainability, this chapter makes the first steps towards synthesizing the three interrelated paradigms: the Effortless Paradigm, the Ecological Principle, and Self-Organization. Together, they possess the qualities inherent in the nature of sustainability itself—adaptability; balance; harmony; and a recognition of interconnectedness.

This unified paradigm is calling for a shift since it is predicated on the pursuit of synergy over subjugation, signaling a shift from AI that seeks to dominate natural and societal systems to one that seeks harmonious integration. Daoism's principle of 'wu-wei', signifying unforced action, attuned to the natural world. Philosopher Martin Buber interprets "non-doing," as a noncoercive and responsive doing, and it is claimed to be the key experience and conception that Europe can learn from Chinese philosophy in order to temper its thirst for power and domination over things and over others (Nelson, E.S et. al, 2014).

Self-organization becomes a metaphor for AI development that seeks synergy rather than subjugation. The ecological concepts become blueprints for AI systems that enhance our ecosystems, championing a development ethos that values coexistence and mutual benefit. The synergy of these paradigms may lead to emerging outcomes where AI is developed not as a forceful tool of human ambition but as an effortless, integrated, and responsible participant in the broader ecological and societal context.

The applications of this new paradigm within AI are multifaceted, offering a spectrum of possibilities for the future direction of the field. Herein, we outline initial steps that exemplify the paths AI development can take, aligning with our approach. The Effortless Paradigm invites us to reimagine AI as a system that operates with a sense of ease and unforced functionality. Such interpretation could be promoting incremental learning, which stands out within machine learning for its ability to continuously assimilate new knowledge from incoming data without the necessity of revisiting the entire original dataset (Luo, Y. et. al, 2020). This characteristic allows for modifications and improvements to be made to the model in a dynamic, ongoing manner, circumventing the exhaustive process typically associated with training a completely new model from scratch. By embracing incremental learning, we enable AI systems to grow and evolve efficiently, mirroring the adaptability of living organisms. This adaptive learning process ensures that AI applications remain relevant and effective within the dynamic contexts in which they operate.

The Ecological Principle broadens our understanding of AI's role within the web of life; one interpretation may encourage AI designers and developers to undertake a mapping of the stakeholders relevant to their task. This initial step is crucial in cultivating awareness of the AI's contextual ecosystem and promoting a conscientious approach to its deployment. By identifying and understanding the various parties affected—ranging from direct users to indirect ecological and social entities—developers can anticipate and mitigate potential adverse impacts; thus fostering; for example; balance when obtaining fair data representations (Pagano, T.P. et. al, 2023). Self-Organization champions, for example, decentralized AI in a fleet of autonomous vehicles; decentralization enables individual cars to make real-time decisions based on immediate environmental inputs and intra-fleet communication (Zhang, Y. et. al, 2021). This not only optimizes traffic flow and reduces congestion, but also reduces central control in the system, a desirable value.

Within the context of our proposed paradigm, which emphasizes harmony, adaptability, and less intervention, transfer learning (Weiss, K. et. al, 2016) may emerge as a favored approach. This technique aligns well with the ethos of reducing unnecessary computational strain and resource consumption, as it leverages pre-trained models and adapts them to new but related tasks, avoiding the need to build and train models from the ground up for every unique application. Operationalizing transfer learning within AI systems means embracing the interconnectedness of knowledge domains. It allows us to draw from a vast pool of pre-existing models (such as models on musical preferences), fine-tuning them with a specific, often smaller set of data that are relevant to the new context (for example, film preferences).

6. CONCLUSION

To sum up, we understand that sustainability is more than just a collection of words or a set of rules. Rather, it is a hope for a more balanced and harmonious world. Our proposed AI development paradigm reflects this vision by placing an emphasis on flexibility, equilibrium, harmony, and a deep understanding of how everything is interrelated. Because it is an evolving and responsive notion, this paradigm is amenable to integration with cutting-edge tech.

Through our research, we have come up with a preliminary proposal for a synthesis of the Effortless Paradigm, the Ecological Principle, and the idea of Self-Organization. This will pave the way for artificial intelligence development that is in harmony with nature and follows the guidelines of ecological integrity and ethical stewardship. This alternative view of artificial intelligence sees technology more as a tool for fostering ecological balance than as a threat to it.

We must exercise prudence in our haste to embrace AI as a panacea for sustainability issues. We need to be careful not to fall into the trap of thinking that AI is the silver bullet for every problem; if we do, we risk ignoring the moral and environmental considerations that are essential to long-term sustainability. Therefore, we advocate for intelligent and prudent AI applications, stressing that their use should go beyond convenience and genuinely conform to the multifaceted aspects of sustainability.

There is a danger that the present focus on measurements to assess AI's impact on the environment would reduce nature to economic measures, ignoring the complexity and worth of natural ecosystems. That is why we're advocating for a change in thinking, away from artificial intelligence (AI) that aims to control social and environmental systems and towards AI that aims to integrate them harmoniously. Given the complexity of our current and future problems, it is quite evident that such an approach is necessary. We can only build a future that respects human ingenuity and the ecological processes that keep life going if we embrace a holistic perspective that combines innovations in technology with the goal of harmony.

REFERENCES

Acevedo, B., & Malevicius, R. (2021). Rethinking Education for Sustainability in Management Education: Going beyond Metrics toward Human Virtues. In *Handbook of Research on International Business and Models for Global Purpose-Driven Companies* (pp. 219–234). IGI Global. DOI: 10.4018/978-1-7998-4909-4.ch012

Asheim, G. B. Sustainability: Ethical Foundations and Economic Properties; Policy Research Working Paper Series 1302; The World Bank: Washington, DC, USA, 1994.

Bachmann, N., Tripathi, S., Brunner, M., & Jodlbauer, H. (2022). The contribution of data-driven technologies in achieving the sustainable development goals. *Sustainability (Basel)*, *14*(5), 2497. DOI: 10.3390/su14052497

Barroso, L. A., & Hölzle, U. (2007). The Case for Energy-Proportional Computing. *Computer*, *40*(12), 33–37. DOI: 10.1109/MC.2007.443

Bender, E. M., Gebru, T., McMillan-Major, A., & Shmitchell, S. On the Dangers of Stochastic Parrots: Can Language Models Be Too Big? In *Proceedings of the 2021 ACM Conference on Fairness, Accountability, and Transparency, Virtual Event*, 3–10 March 2021; pp. 610–623. DOI: 10.1145/3442188.3445922

Bengio, Y., Delalleau, O., & Le Roux, N. (2005). The Curse of Dimensionality for Local Kernel Machines. *Techn. Rep.*, *1258*, 1.

Berti-Equille, L., Dao, D., Ermon, S., & Goswami, B. Challenges in KDD and ML for Sustainable Development. In *Proceedings of the 27th ACM SIGKDD Conference on Knowledge Discovery & Data Mining*, Singapore, 14 August 2021; pp. 4031–4032. DOI: 10.1145/3447548.3470798

Bexell, M., & Jönsson, K. (2017). Responsibility and the United Nations' sustainable development goals. *Forum for Development Studies*, *44*(1), 13–29. DOI: 10.1080/08039410.2016.1252424

Biermann, F., Kanie, N., & Kim, R. E. (2017). Global governance by goal-setting: The novel approach of the UN sustainable development goals. *Current Opinion in Environmental Sustainability*, *26–27*, 26–31. DOI: 10.1016/j.cosust.2017.01.010

Bolte, L., Vandemeulebroucke, T., & Van Wynsberghe, A. (2022). From an Ethics of Carefulness to an Ethics of Desirability: Going Beyond Current Ethics Approaches to Sustainable AI. *Sustainability (Basel)*, *14*(8), 4472. DOI: 10.3390/su14084472

Brevini, B. (2021). *Is AI Good for the Planet?* Polity Press. [Google Scholar]

Brown, R. E., Brown, R., Masanet, E., Nordman, B., Tschudi, B., Shehabi, A., Stanley, J., Koomey, J., Sartor, D., & Chan, P. (2007). *Report to Congress on Server and Data Center Energy Efficiency: Public Law 109-431; Lawrence Berkeley National Laboratory*. LBNL.

Bullard, R. D. (1993). *Confronting Environmental Racism: Voices from the Grassroots*. South End Press.

Cao, T., Lim, E.-P., Zhou, Z.-H., Ho, T.-B., Cheung, D., & Motoda, H. (Eds.). Springer International Publishing: Cham, Switzerland, 2015; Volume 9078, pp. 27–38. ISBN 978-3-319-18031-1.

Coeckelbergh, M. (2021). AI for Climate: Freedom, Justice, and Other Ethical and Political Challenges. *AI and Ethics*, *1*(1), 67–72. DOI: 10.1007/s43681-020-00007-2

Coulson, R. N., Folse, L. J., & Loh, D. K. (1987). Artificial intelligence and natural resource management. *Science*, *237*(4812), 262–267. DOI: 10.1126/science.237.4812.262 PMID: 17772054

Cowls, J., Tsamados, A., Taddeo, M., & Floridi, L. (2023). The AI Gambit: Leveraging Artificial Intelligence to Combat Climate Change—Opportunities, Challenges, and Recommendations. *AI & Society*, *38*(1), 283–307. DOI: 10.1007/s00146-021-01294-x PMID: 34690449

Crawford, K. (2021). *The Atlas of AI: Power, Politics, and the Planetary Costs of Artificial Intelligence*. Yale University Press.

Cui, S., Gao, Y., Huang, Y., Shen, L., Zhao, Q., Pan, Y., & Zhuang, S. (2023). Advances and Applications of Machine Learning and Deep Learning in Environmental Ecology and Health. *Environmental Pollution, 335*, 122358. DOI: 10.1016/j.envpol.2023.122358 PMID: 37567408

Dhar, P. (2020). The carbon impact of artificial intelligence. *Nature Machine Intelligence*, *2*(8), 423–425. DOI: 10.1038/s42256-020-0219-9

Dilkina, B., Houtman, R., Gomes, C. P., Montgomery, C. A., McKelvey, K. S., Kendall, K., Graves, T. A., Bernstein, R., & Schwartz, M. K. (2017). Trade-offs and Efficiencies in Optimal Budget-constrained Multispecies Corridor Networks. [Google Scholar] [CrossRef]. *Conservation Biology*, *31*(1), 192–202. DOI: 10.1111/cobi.12814 PMID: 27677418

Fisher, D. A Selected Summary of AI for Computational Sustainability. In *Proceedings of the Thirty-First AAAI Conference on Artificial Intelligence (AAAI-17)*, San Francisco, CA, USA, 4–9 February 2017. [Google Scholar] DOI: 10.1609/aaai.v31i1.11148

Ganaie, M. A., Hu, M., Malik, A. K., Tanveer, M., & Suganthan, P. N. (2022). Ensemble Deep Learning: A Review. *Engineering Applications of Artificial Intelligence*, *115*, 105151. DOI: 10.1016/j.engappai.2022.105151

Garbin, C., Zhu, X., & Marques, O. (2020). Dropout vs. Batch Normalization: An Empirical Study of Their Impact to Deep Learning. *Multimedia Tools and Applications*, *79*(19-20), 12777–12815. DOI: 10.1007/s11042-019-08453-9

Haefner, N., Wincent, J., Parida, V., & Gassmann, O. (2021). Artificial intelligence and innovation management: A review, framework, and research agenda . *Technological Forecasting and Social Change*, *162*, 120392. DOI: 10.1016/j.techfore.2020.120392

Hanin, B. Which Neural Net Architectures Give Rise to Exploding and Vanishing Gradients? In *Proceedings of the 32nd Conference on Neural Information Processing Systems (NeurIPS 2018)*, Montréal, Canada, 2–8 December 2018; Volume 31.

Heilinger, J.-C., Kempt, H., & Nagel, S. (2023). Beware of Sustainable AI! Uses and Abuses of a Worthy Goal. [Google Scholar] [CrossRef]. *AI and Ethics*, *4*, 1–12.

Hsieh, C.-J., Si, S., & Dhillon, I. A Divide-and-Conquer Solver for Kernel Support Vector Machines. In Proceedings of the 31st International Conference on Machine Learning, PMLR, Beijing, China, 22–24 June 2014; pp. 566–574.

Kaack, L. H., Donti, P. L., Strubell, E., Kamiya, G., Creutzig, F., & Rolnick, D. (2022). Aligning artificial intelligence with climate change mitigation. *Nature Climate Change*, *12*(6), 518–527. DOI: 10.1038/s41558-022-01377-7

Kemp, R., & Martens, P. (2007). Sustainable Development: How to Manage Something That Is Subjective and Never Can Be Achieved? *Sustainability*, *3*(2), 5–14. DOI: 10.1080/15487733.2007.11907997

Lansing, J. S., & Kremer, J. N. (1993). Emergent Properties of Balinese Water Temple Networks: Coadaptation on a Rugged Fitness Landscape. *American Anthropologist*, *95*(1), 97–114. DOI: 10.1525/aa.1993.95.1.02a00050

Loy, D. (1985). Wei-Wu-Wei: Nondual Action. *Philosophy East & West*, *35*(1), 73–86. DOI: 10.2307/1398682

Luo, Q. Daoism and Environmental Sustainability—A Completely Different Way of Thinking. In *Proceedings of the International Workshop on Sustainable City Region*, Bali, Indonesia, 23–24 February 2009; pp. 164–171.

Luo, Y., Yin, L., Bai, W., & Mao, K. (2020). An Appraisal of Incremental Learning Methods. *Entropy (Basel, Switzerland)*, *22*(11), 1190. DOI: 10.3390/e22111190 PMID: 33286958

Mensah, J. (2019). Mensah, J. Sustainable Development: Meaning, History, Principles, Pillars, and Implications for Human Action: Literature Review. *Cogent Social Sciences*, *5*(1), 1653531. DOI: 10.1080/23311886.2019.1653531

Nagler, J., van den Hoven, J., & Helbing, D. (2019). An Extension of Asimov's Robotics Laws. In Helbing, D. (Ed.), *Towards Digital Enlightenment*. Springer. DOI: 10.1007/978-3-319-90869-4_5

Nash, K. L., Blythe, J. L., Cvitanovic, C., Fulton, E. A., Halpern, B. S., Milner-Gulland, E., Addison, P. F., Pecl, G. T., Watson, R. A., & Blanchard, J. L. (2020). To Achieve a Sustainable Blue Future, Progress Assessments Must Include Interdependencies between the Sustainable Development Goals. *One Earth*, *2*(2), 161–173. DOI: 10.1016/j.oneear.2020.01.008

Nelson, E. S. (2014). Technology and the Way: Buber, Heidegger, and Lao-Zhuang "Daoism". *Journal of Chinese Philosophy*, *41*(3-4), 307–327. DOI: 10.1163/15406253-0410304005

Nelson, E. S. (2020). *Daoism and Environmental Philosophy: Nourishing Life*. Routledge. DOI: 10.4324/9780429399145

Nguyen, A.-P., Moreno, D. L., Le-Bel, N., & Rodríguez Martínez, M. (2023). MonoNet: Enhancing Interpretability in Neural Networks via Monotonic Features. *Bioinformatics Advances*, *3*(1), vbad016. DOI: 10.1093/bioadv/vbad016 PMID: 37143924

Nordgren, A. (2023). Artificial Intelligence and Climate Change: Ethical Issues. *J. Inf. Commun. Ethics Soc.*, *21*(1), 1–15. DOI: 10.1108/JICES-11-2021-0106

Ofek, N., Katz, G., Shapira, B., & Bar-Zev, Y. Sentiment Analysis in Transcribed Utterances. In Advances in Knowledge Discovery and Data Mining; Lecture Notes in Computer Science; 2020.

Ofek, N., Rokach, L., Stern, R., & Shabtai, A. (2017). Fast-CBUS: A Fast Clustering-Based Undersampling Method for Addressing the Class Imbalance Problem. *Neurocomputing*, *243*, 88–102. DOI: 10.1016/j.neucom.2017.03.011

Pagano, T. P., Loureiro, R. B., Lisboa, F. V. N., Peixoto, R. M., Guimarães, G. A. S., Cruz, G. O. R., Araujo, M. M., Santos, L. L., Cruz, M. A. S., Oliveira, E. L. S., Winkler, I., & Nascimento, E. G. S. (2023). Bias and Unfairness in Machine Learning Models: A Systematic Review on Datasets, Tools, Fairness Metrics, and Identification and Mitigation Methods. *Big Data and Cognitive Computing*, *7*(1), 15. DOI: 10.3390/bdcc7010015

. Patterson, D.; Gonzalez, J.; Le, Q.; Liang, C.; Munguia, L.-M.; Rothchild, D.; So, D.; Texier, M.; Dean, J. Carbon Emissions and Large Neural Network Training. arXiv 2021,

Pierce, J., & Jameton, A. (2004). *The Ethics of Environmentally Responsible Health Care*. Oxford University Press.

Portier, K., Greer, G. E., Rokach, L., Ofek, N., Wang, Y., Biyani, P., Yu, M., Banerjee, S., Zhao, K., Mitra, P., & Yen, J. (2013). Understanding Topics and Sentiment in an Online Cancer Survivor Community. *Journal of the National Cancer Institute. Monographs*, *2013*(47), 195–198. DOI: 10.1093/jncimonographs/lgt025 PMID: 24395991

Provost, F., & Kolluri, V. (1999). A Survey of Methods for Scaling Up Inductive Algorithms. *Data Mining and Knowledge Discovery*, *3*(2), 131–169. DOI: 10.1023/A:1009876119989

Report: Inequalities Exacerbate Climate Impacts on Poor. Available online: https://www.un.org/sustainabledevelopment/blog/2016/10/report-inequalities-exacerbate-climate-impacts-on-poor/(accessed on 24 September 2023).

Salvia, A. L., Leal Filho, W., Brandli, L. L., & Griebeler, J. S. (2019). Assessing research trends related to sustainable development goals: Local and global issues. *Journal of Cleaner Production*, *208*, 841–849. DOI: 10.1016/j.jclepro.2018.09.242

Schwartz, R., Dodge, J., Smith, N. A., Etzioni, O., & Green, A. I. (2020). Green AI. *Communications of the ACM*, *63*(12), 54–63. DOI: 10.1145/3381831

Seeley, T. D. (2002). When Is Self-Organization Used in Biological Systems? *The Biological Bulletin*, *202*(3), 314–318. DOI: 10.2307/1543484 PMID: 12087005

Sison, A. J. G., Daza, M. T., Gozalo-Brizuela, R., & Garrido-Merchán, E. C. ChatGPT: More than a Weapon of Mass Deception, Ethical Challenges and Responses from the Human-Centered Artificial Intelligence (HCAI) Perspective. arXiv 2023, arXiv:2304.11215. [Google Scholar]

Van Wynsberghe, A., & Sustainable, A. I. (2021). AI for Sustainability and the Sustainability of AI. *AI and Ethics*, *1*(3), 213–218. DOI: 10.1007/s43681-021-00043-6

Vinuesa, R., Azizpour, H., Leite, I., Balaam, M., Dignum, V., Domisch, S., Felländer, A., Langhans, S. D., Tegmark, M., & Fuso Nerini, F. (2020). The Role of Artificial Intelligence in Achieving the Sustainable Development Goals. *Nature Communications*, *11*(1), 233. DOI: 10.1038/s41467-019-14108-y PMID: 31932590

Weiss, K., Khoshgoftaar, T. M., & Wang, D. (2016). A Survey of Transfer Learning. *Journal of Big Data*, *3*(1), 9. DOI: 10.1186/s40537-016-0043-6

Wu, X., Gomes-Selman, J., Shi, Q., Xue, Y., Garcia-Villacorta, R., Anderson, E., Sethi, S., Steinschneider, S., Flecker, A., & Gomes, C. Efficiently Approximating the Pareto Frontier: Hydropower Dam Placement in the Amazon Basin. In *Proceedings of the AAAI Conference on Artificial Intelligence*, New Orleans, LA, USA, 2–7 February 2018; Volume 32. DOI: 10.1609/aaai.v32i1.11347

Yang, T.-J., Chen, Y.-H., Emer, J., & Sze, V. A Method to Estimate the Energy Consumption of Deep Neural Networks. In *Proceedings of the 51st Asilomar Conference on Signals, Systems, and Computers*, Pacific Grove, CA, USA, 29 October–1 November 2017; IEEE: Piscataway, NJ, USA, 2017; pp. 1916–1920 DOI: 10.1109/ACSSC.2017.8335698

Zhang, Y., Macke, W., Cui, J., Urieli, D., & Stone, P. (2021). Learning a Robust Multiagent Driving Policy for Traffic Congestion Reduction. *Neural Computing & Applications*, •••, 1–14.

Zhou, X., Zhang, W., Chen, Z., Diao, S., & Zhang, T. (2021). Efficient Neural Network Training via Forward and Backward Propagation Sparsification. *Advances in Neural Information Processing Systems*, *34*, 15216–15229.

Chapter 8
Bibliometrics of Artificial Intelligence and Economy

Frida Sanchez Saldivar
Autonomous University of Tamaulipas, Mexico

Mayra Elizabeth Brosig Rodriguez
https://orcid.org/0000-0002-0472-5266
Universidad Autonoma de Nuevo Leon, Mexico

Kenia Garza
Autonomous University of Tamaulipas, Mexico

ABSTRACT

This study presents a systematic analysis of the scientific literature on the relationship between artificial intelligence (AI) and the economy during the period 2020-2024, based on the analysis of 530 documents indexed in Web of Science. Employing a methodology that combines quantitative analyses of scientific production, co-citation networks, and keyword mapping using R software and VOSviewer, the research identifies the main trends, collaboration patterns, and emerging areas in this field. The results demonstrate sustained growth in scientific production, highlighting a significant concentration in three main areas: business digital transformation, supply chain management, and innovation in business models. The

DOI: 10.4018/979-8-3693-8643-9.ch008

fundamental contribution of this study lies in providing an updated and systematic understanding of the state of research in AI and economics, offering clear directions for future research and practical guidance for professionals in the field.

INTRODUCTION

Artificial intelligence (AI) has become a transformative force that is redefining economic and business paradigms in the 21st century (Cantú-Ortíz et al., 2023). This technology, which ranges from machine learning to deep learning, is generating significant changes in various economic sectors (Janiesch et al., 2021), modifying traditional business models and creating new opportunities for growth and development (Burström et al., 2021). The rapidity with which AI is evolving and its increasing adoption in different economic contexts necessitates a systematic analysis of the scientific literature to understand the current state of research in this field.

The intersection between AI and the economy has generated growing academic interest, as demonstrated by the significant volume of scientific publications in recent years (Qin et al., 2024). This interest is particularly reflected in areas such as supply chain management (Wamba & Queiroz, 2020), digital transformation of businesses (Guo et al., 2023), and innovation in business models (Burström et al., 2021). This suggests that AI is not only improving operational efficiency but also creating new paradigms in the way economic activities are conceived and executed (Waltersmann et al., 2021; Lu, 2021).

However, there is a need to systematize and comprehensively analyze the different perspectives and findings that have been generated in recent literature. Understanding the trends, patterns, and areas of emphasis in research on AI and economics is fundamental to identifying knowledge gaps and guiding future research.

The present study aims to conduct a systematic bibliometric analysis of the scientific literature on the relationship between artificial intelligence and the economy, published between 2020 and 2024. This analysis seeks to:

1. Identify the main trends and patterns in research on AI and economics
2. Examine the geographical distribution of scientific production in this field
3. Analyze collaboration and citation networks among researchers
4. Determine the predominant and emerging thematic areas
5. Identify possible gaps in the current literature

This work is based on the analysis of scientific documents indexed in Web of Science, providing a comprehensive view of the current state of research at the intersection of AI and economics. The methodology employed allows not only to quantify scientific production but also to identify the interrelationships between different concepts and research areas.

The results of this systematic analysis will contribute to a better understanding of the current landscape of research in AI and economics, providing a solid foundation for future studies and helping to identify potential research areas that require greater attention.

METHODOLOGY

This study employs a systematic bibliometric analysis to examine the scientific production at the intersection of artificial intelligence and economics. The methodology was structured in several phases to ensure a comprehensive analysis of the existing literature, relying on specialized tools for the processing and analysis of bibliometric data.

Data collection was conducted through the Web of Science (WoS) platform, recognized for its comprehensive coverage and rigor in indexing scientific publications. Data were extracted in .bib format directly from the WoS platform to ensure the integrity of the bibliometric information. For the processing and statistical analysis of the data, R software was used, leveraging its advanced capabilities for analyzing and visualizing scientific data.

The search was structured considering the following search terms within the platform: ((“artificial intelligence” OR “AI” OR “machine learning” OR “deep learning”) AND (“econom” OR “business” OR “market” OR "industry" OR "financial")) AND ("impact" OR "transform*" OR "re-shap*" OR "disrupt*" OR "adopt*" OR "implement*") AND ("growth" OR "productivity" OR "labor market" OR "job market" OR "business model*")).* This search strategy was applied to the period 2020-2024 to capture the most recent trends in the field, resulting in the initial identification of 530 documents.

The analysis of the sample allowed for the classification of documents according to their typology, identifying 433 research articles, 60 review articles, 36 early access articles, and 1 book review. The processing of these data using R statistical software facilitated the generation of robust statistical analyses and comprehensive visualizations of trends and patterns in the literature. The use of this statistical software, in conjunction with the data extracted in .bib format from WoS, enabled detailed co-citation analyses, mapping of collaboration networks, and analysis of thematic trends with a high level of precision and reliability.

RESULTS

The analysis of annual scientific production reveals a dynamic trend in research on AI and economics during the period 2020-2024. The data suggest significant growth in the number of publications, indicating increasing academic interest in the intersection between artificial intelligence and economic aspects. This sustained increase in scientific production reflects the growing relevance of AI in economic and business contexts, as well as the maturation of this field of research.

Figure 1. Annual scientific production

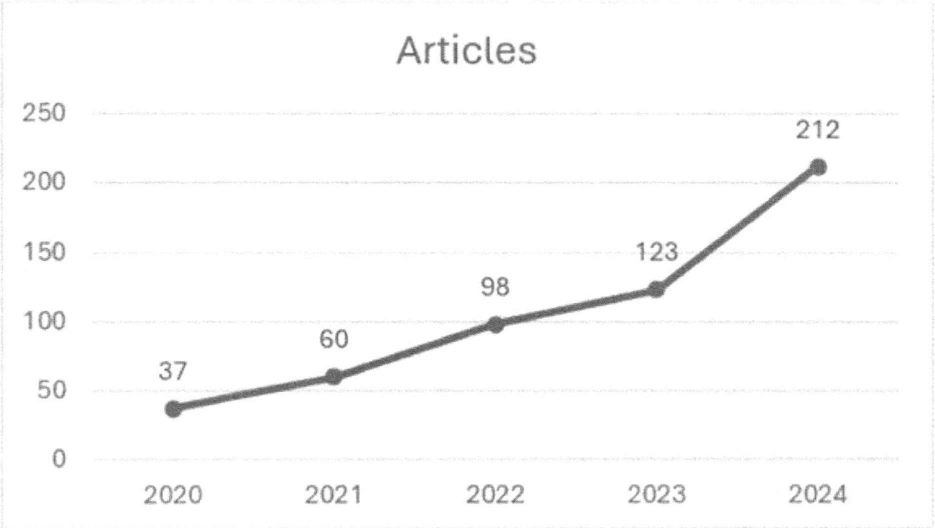

Source: Elaboration from integrated database of Web of Science platform

Most Relevant Journals In The Study Of The Topic

The analysis of the most relevant journals in the field shows a diversified distribution of scientific production. Publications are concentrated in specialized journals covering both technological and economic aspects, which underscores the interdisciplinary nature of this research area. This distribution suggests that the impact of AI on the economy is being studied from multiple academic perspectives, enriching the debate and understanding of the phenomenon..

Figure 2. Most Relevant Journals

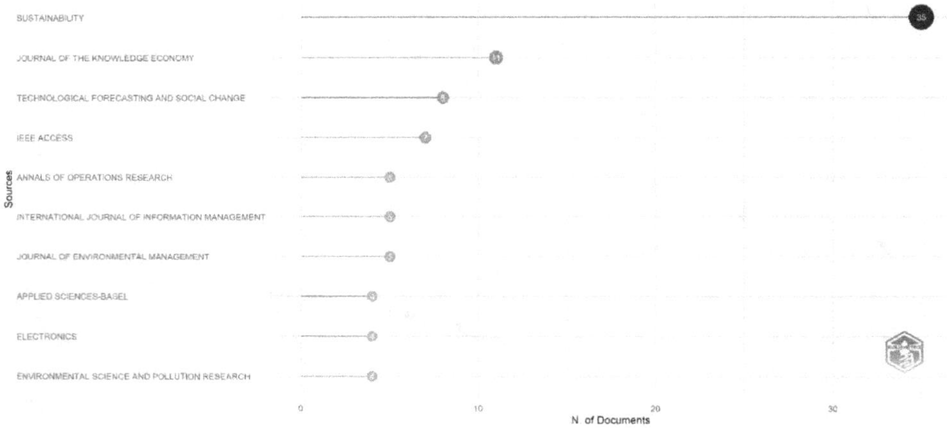

Source: Web of Science Platform

Corresponding Author Country

The geographical distribution of corresponding authors reveals interesting patterns in global scientific production. The data show a significant concentration in certain countries, which could indicate the existence of centers of excellence in research on AI and economics. This geographical distribution also reflects the different research capabilities and resources available in different regions of the world, as well as national priorities in terms of AI research and development.

Figure 3. Corresponding author country

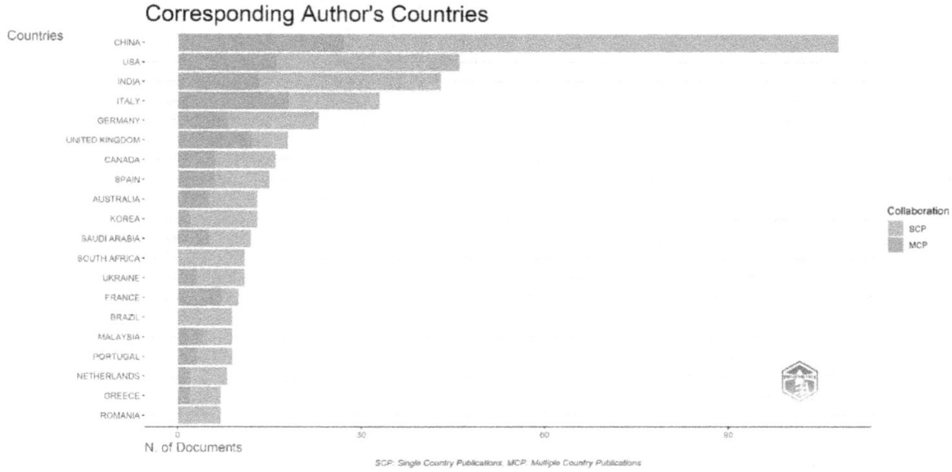

Source: Web of Science Platform

Countries With The Highest Number Of Citations

The analysis of citations by country reveals the influence and international impact of research in this field. It is observed that certain countries emerge as leaders not only in terms of scientific production but also in the influence of their publications, measured by the number of citations received. This metric provides a perspective on the quality and impact of research beyond the quantity of publications, identifying the main centers of influence in the generation of knowledge on the subject.

Figure 4. Most cited countries

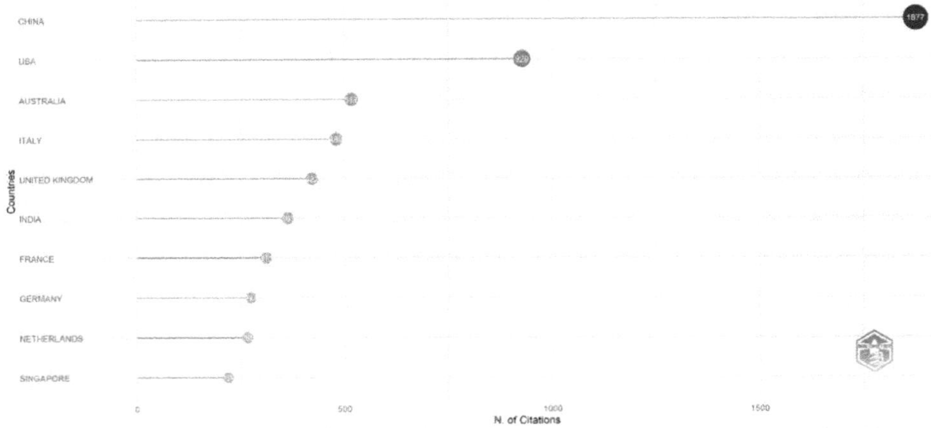

Source: Web of Science Platform

Most Cited Documents

Table 1 presents the ten most cited documents within the study topic during the period 2020-2024. The analysis of these works reveals significant patterns in the evolution and impact of AI in various economic sectors. The most cited article by Wamba (245 citations) addresses the integration of blockchain in supply chain management, reflecting the growing importance of distributed technologies in conjunction with AI for the optimization of business processes. It is notable that the most influential works cover a wide range of applications, from agriculture (Khana, 199 citations) to the digital transformation of business models (Burstrom, 148 citations), demonstrating the cross-cutting nature of AI's impact on the economy.

Table 1. Most cited documents

Author	Article	Number of citations
Wamba S. (2020)	*Blockchain in the operations and supply chain management: Benefits, challenges and future research opportunities*	245
Khana S. (2020)	*Remote Sensing in Agriculture—Accomplishments, Limitations, and Opportunities*	199
Ciarli T. (2021)	*Digital technologies, innovation, and skills: Emerging trajectories and challenges*	198
Cai Y. (2020)	*Omni-channel management in the new retailing era: A systematic review and future research agenda*	190
Sepasgozar S. (2021)	*Differentiating Digital Twin from Digital Shadow: Elucidating a Paradigm Shift to Expedite a Smart, Sustainable Built Environment*	184
Yag, I. (2022)	*Artificial Intelligence-Based Robust Hybrid Algorithm Design and Implementation for Real-Time Detection of Plant Diseases in Agricultural Environments*	178
Attaran M. (2020)	*Digital technology enablers and their implications for supply chain management*	169
Trenerry B. (2021)	*Preparing Workplaces for Digital Transformation: An Integrative Review and Framework of Multi-Level Factors*	154
Burstrom. T. (2021)	*AI-enabled business-model innovation and transformation in industrial ecosystems: A framework, model and outline for further research*	148
Guo X. (2023)	*Does digital transformation improve the firm's performance? From the perspective of digitalization paradox and managerial myopia*	144

Source: Elaboration from the database of the Web of Science platform

The temporal distribution of these highly cited publications, with a significant concentration in the years 2020-2021, suggests that these years were particularly important for establishing the theoretical and practical foundations of AI integration in various economic sectors. It is especially relevant to observe that more recent works, such as Guo (2023), have already accumulated a significant number of citations (144), indicating a sustained and growing interest in understanding the effects of digital transformation and AI on business performance. This trend in citations reflects the maturation of the field and the consolidation of lines of research connecting AI with different aspects of the economy and business.

Most Frequent Words

The analysis of word frequency reveals significant patterns. The distribution of frequent terms shows a clear predominance of concepts related to digital transformation and its economic implications. There is a high frequency of technical terms specific to the AI field, such as "machine learning" and "artificial intelligence," along with economic terms such as "performance," "innovation," and "business," reflecting the close integration between technology and business aspects in the current literature.

Figure 5. Most frecuent keywords

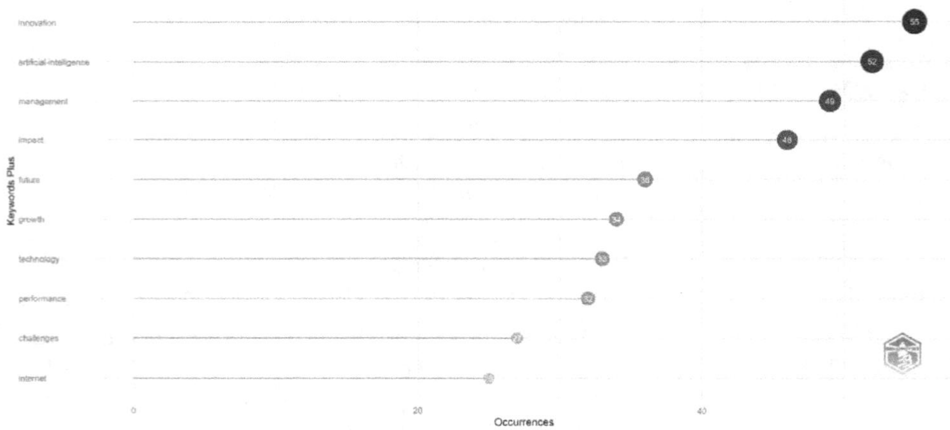

Source: Web of Science Platform

Keywords Clouds

The keyword cloud (Figure 6) provides an intuitive visualization of the central concepts that dominate the research field, revealing through the arrangement and size of the terms several significant thematic clusters.

Figure 6. Keyword cloud

Source: Web of Science Platform

The technological cluster emerges as a fundamental component, dominated by terms such as "artificial intelligence," "machine learning," and "deep learning," indicating the centrality of these technologies in current research. In parallel, a robust economic cluster is identified, characterized by terms such as "performance," "innovation," "strategy," and "business model," which reflect the marked focus on the business implications of these technologies. The visualization also reveals a significant transformation cluster, evidenced by terms such as "digital transformation," "adoption," and "implementation," which signal the importance of organizational change processes in the field of study. Complementing these groups, an impact cluster is distinguished, represented by terms such as "sustainability," "efficiency," and "optimization," indicating the objectives and expected results of the implementation of these technologies in economic and business contexts.

Trending Topics

The analysis of trending topics identified in the research (Figure 7) reveals a dynamic evolution of the field and highlights significant emerging areas at the intersection of artificial intelligence and economics. Among the main trends, digital transformation is positioned as the dominant theme, with particular emphasis on aspects of implementation and management of organizational change. This trend is accompanied by a growing interest in business model innovation, where sustained attention is observed towards the adaptation of traditional business structures to new technological paradigms. Likewise, the connection between AI and sustainability emerges as a significant trend, reflecting a growing awareness of the role of technology in achieving sustainable development goals.

Regarding emerging trends, a growing interest in intelligent automation is identified, specifically in its application to business processes, suggesting a deepening in the integration of AI into the daily operations of organizations. At the same time, there is an increase in attention paid to the ethical and governance aspects of AI in economic contexts, reflecting a maturation of the field towards broader considerations than purely technical ones. Digital resilience also stands out as an emerging trend, with a growing emphasis on the development of organizational capabilities to adapt and respond effectively to changes through the use of AI-based technologies.

Figure 7. Trending topics

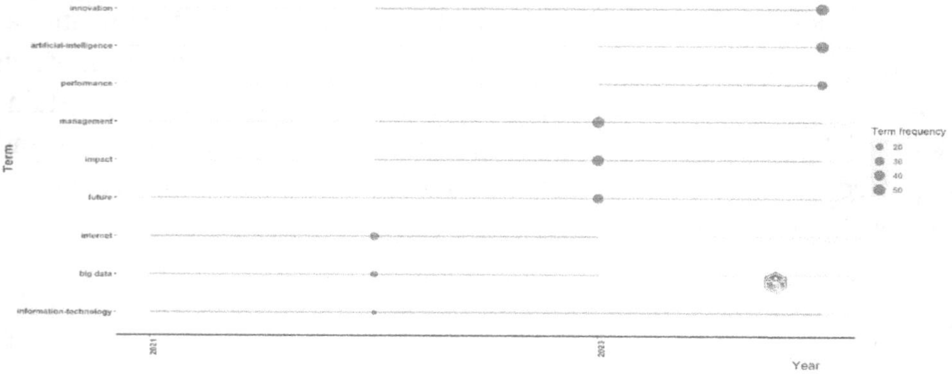

Source: Web of Science Platform

Keyword Network

The analysis of the co-occurrence network of keywords, carried out using VOSviewer software, reveals a complex and multifaceted structure of relationships between the main concepts in the field of AI and economics. At the center of the network, a dominant cluster is identified organized around terms such as "artificial intelligence" and "machine learning," which act as the conceptual core of the field. This central cluster shows strong connections with practical applications and implementation methodologies, suggesting its fundamental role as a bridge between different research areas.

Figure 8. A network cluster centered around artificial intelligence

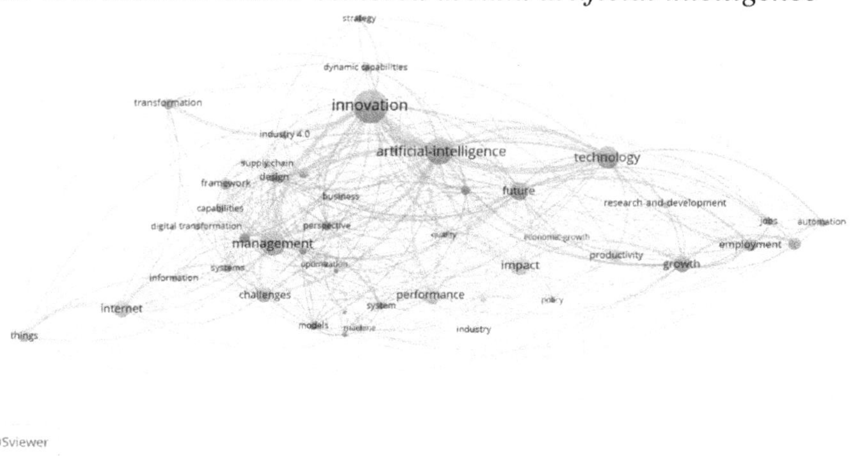

The network visualization also highlights a significant cluster centered on business transformation, where terms such as "digital transformation," "innovation," and "business model" show a strong interconnection with both technological and organizational aspects. This connection structure emphasizes the role of AI as a fundamental catalyst in the transformation of traditional business models. In parallel, a robust cluster related to economic impact is observed, grouping terms such as "performance," "productivity," and "economic growth," evidencing the direct links between AI adoption and measurable economic outcomes.

A particularly relevant finding is the emergence of a distinctive cluster that connects AI with sustainability objectives, reflecting the growing importance of environmental and social considerations in the implementation of AI technologies. The connectivity patterns in the network reveal the densest connections between technological and business clusters, suggesting a deep integration between these domains. The center-periphery structure of the network is notable, with terms related to AI and digital transformation showing high centrality, while more specific or emerging concepts are located in the periphery, indicating a clear hierarchy in the existing literature.

The spatial arrangement of clusters and their interconnections suggests an evolution of the field from purely technical aspects to broader impact considerations, with new connections emerging especially in sustainability and ethics themes. This structure reveals the maturation of the field towards a more integrated and multidisciplinary perspective. The conceptual bridges identified in the network point to significant opportunities for interdisciplinary research and suggest potential areas for new investigation. Peripheral connections, in particular, indicate emerging directions in research that could develop in the near future.

This network analysis ultimately provides a comprehensive view of the knowledge structure in the field of AI and economics, revealing both established patterns and emerging directions in research, and suggesting promising paths for future research in this rapidly evolving area.

CONCLUSION

The present analysis of literature on artificial intelligence and economics reveals significant patterns and emerging trends that characterize the evolution of this research field during the period 2020-2024. The results obtained allow us to draw several relevant conclusions about the current state and future directions of research in this area.

First, the analysis of scientific production demonstrates sustained growth in the number of publications, with 530 documents identified, of which 433 are research articles. This distribution suggests an actively expanding research field, with particular emphasis on the generation of new empirical knowledge. The concentration of publications in high-impact journals indicates the relevance and maturity that this field of study is reaching.

The citation analysis reveals a structure of academic influence where fundamental works stand out, such as Wamba (2020) with 245 citations, which has established important foundations in understanding the integration of emerging technologies in business management. The geographical distribution of scientific production shows a significant concentration in certain countries, suggesting the existence of centers of excellence in research on AI and economics, although it also points to opportunities to expand international participation in this field.

The analysis of co-occurrence of keywords and the identification of thematic clusters reveal an evolution from purely technical perspectives towards more integrated approaches that consider organizational, social, and environmental aspects. The emergence of clusters related to sustainability and ethics suggests a maturation of the field towards more holistic considerations of the impact of AI on the economy.

The identified trends point towards three main directions of development: digital transformation as a central axis of organizational change, innovation in business models as an area of practical application, and the integration of sustainability considerations in the implementation of AI-based solutions. These trends suggest that the field is evolving towards a more sophisticated understanding of the interaction between technology and economic systems.

A particularly relevant conclusion is the identification of significant gaps in the current literature, especially in areas such as the assessment of the long-term impact of AI on specific labor markets, the precise quantification of return on investment in AI implementations, and the development of regulatory frameworks adapted to new technological realities.

The implications of this study are significant for both future research and practice. For researchers, promising opportunities are identified in the study of the intersection between AI, sustainability, and economic development. For professionals and decision-makers, the results provide a solid basis for understanding current trends and anticipating future developments in the application of AI in economic contexts.

Finally, this bibliometric analysis contributes to the existing literature by providing a systematic and comprehensive view of the current state of research in AI and economics, establishing a solid foundation for future research and pointing to promising directions for the development of the field. It is recommended that future studies delve deeper into the identified emerging areas and address the knowledge gaps indicated in this analysis.

REFERENCES

Attaran, M. (2020). Digital technology enablers and their implications for supply chain management. *Supply Chain Forum: An International Journal (21)*3, 158-172. DOI: 10.1080/16258312.2020.1751568

Burström, T., Parida, V., Lahti, T., & Wincent, J. (2021). AI-enabled business-model innovation and transformation in industrial ecosystems: A framework, model and outline for further research. *Journal of Business Research, 127*, 85–95. DOI: 10.1016/j.jbusres.2021.01.016

Cai, Y. J., & Lo, C. K. (2020). Omni-channel management in the new retailing era: A systematic review and future research agenda. *International Journal of Production Economics, 229*, 107729. DOI: 10.1016/j.ijpe.2020.107729

Cantú-Ortiz, F. J., Galeano Sánchez, N., Garrido, L., Terashima-Marin, H., & Brena, R. F. (2020). An artificial intelligence educational strategy for the digital transformation. [IJIDeM]. *International Journal on Interactive Design and Manufacturing, 14*(4), 1195–1209. DOI: 10.1007/s12008-020-00702-8

Ciarli, T., Kenney, M., Massini, S., & Piscitello, L. (2021). Digital technologies, innovation, and skills: Emerging trajectories and challenges. *Research Policy, 50*(7), 104289. DOI: 10.1016/j.respol.2021.104289

Guo, X., Li, M., Wang, Y., & Mardani, A. (2023). Does digital transformation improve the firm's performance? From the perspective of digitalization paradox and managerial myopia. *Journal of Business Research, 163*, 113868. DOI: 10.1016/j.jbusres.2023.113868

Janiesch, C., Zschech, P., & Heinrich, K. (2021). Machine learning and deep learning. *Electronic Markets, 31*(3), 685–695. DOI: 10.1007/s12525-021-00475-2

Khanal, S., Kc, K., Fulton, J. P., Shearer, S., & Ozkan, E. (2020). Remote sensing in agriculture—Accomplishments, limitations, and opportunities. *Remote Sensing (Basel), 12*(22), 3783. DOI: 10.3390/rs12223783

Lu, C. H. (2021). The impact of artificial intelligence on economic growth and welfare. *Journal of Macroeconomics*, *69*, 103342. DOI: 10.1016/j.jmacro.2021.103342

Qin, Y., Xu, Z., Wang, X., & Skare, M. (2024). Artificial intelligence and economic development: An evolutionary investigation and systematic review. *Journal of the Knowledge Economy*, *15*(1), 1736–1770. DOI: 10.1007/s13132-023-01183-2

Sepasgozar, S. M. (2021). Differentiating digital twin from digital shadow: Elucidating a paradigm shift to expedite a smart, sustainable built environment. *Buildings (Basel, Switzerland)*, *11*(4), 151. DOI: 10.3390/buildings11040151

Trenerry, B., Chng, S., Wang, Y., Suhaila, Z. S., Lim, S. S., Lu, H. Y., & Oh, P. H. (2021). Preparing workplaces for digital transformation: An integrative review and framework of multi-level factors. *Frontiers in Psychology*, *12*, 620766. DOI: 10.3389/fpsyg.2021.620766 PMID: 33833714

Waltersmann, L., Kiemel, S., Stuhlsatz, J., Sauer, A., & Miehe, R. (2021). Artificial intelligence applications for increasing resource efficiency in manufacturing companies—A comprehensive review. *Sustainability (Basel)*, *13*(12), 6689. DOI: 10.3390/su13126689

Wamba, S. F., & Queiroz, M. M. (2020). Blockchain in the operations and supply chain management: Benefits, challenges and future research opportunities. *International Journal of Information Management*, *52*, 102064. DOI: 10.1016/j.ijinfomgt.2019.102064

Yağ, İ., & Altan, A. (2022). Artificial intelligence-based robust hybrid algorithm design and implementation for real-time detection of plant diseases in agricultural environments. *Biology (Basel)*, *11*(12), 1732. DOI: 10.3390/biology11121732 PMID: 36552243

Chapter 9
The Use of AI Applications in the Economy

Daniel Avila-Guzman
https://orcid.org/0009-0008-9547-7544
Autonomous University of Tamaulipas, Mexico

Vicente Villanueva
https://orcid.org/0000-0002-9388-5746
Autonomous University of Tamaulipas, Mexico

Juan Carlos Huerta
https://orcid.org/0009-0005-4713-5759
Autonomous University of Tamaulipas, Mexico

Jesus Hernandez Zamorano
Autonomous University of Tamaulipas, Mexico

ABSTRACT

AI in the economy is based on the use and application of technology to improve efficiency, effectiveness, and productivity aimed at optimizing decision making in organizations. It is currently a mechanism or tool that has transformed the way operations are carried out; this has allowed for

DOI: 10.4018/979-8-3693-8643-9.ch009

significant great advances and the results are favorable for organizations in the department or area in which they are applied. Some of the most relevant areas of AI in the economy for decision making are the area of finance, accounting, digital marketing, in each of them what will allow you to know customers and their markets in a more detailed and optimal way, with these tools you can predict, personalize consumer behavior, thanks to the analysis of a large volume of big data information. Nowadays, generative AI processes are models or processes of deep learning, which is a derivative of machine learning, its main function is to simulate and model behaviors of large amounts of data, this allows organizations to make better decisions.

INTRODUCTION

In this chapter, what we intend to illustrate is how AI carries out processes that help decision making.

In the opinion of Kurian N et. a (Kurian N, 2023) defines AI as the ability of machines to simulate the cognitive processes that are typical of the human species, it has become in just over two years a disruptive force in multiple knowledge management sectors

For his part, Boden (Boden, 2018) explains that artificial intelligence (AI) is a scientific and technological discipline that seeks to create systems capable of solving tasks that normally require human intelligence. In recent years, this discipline has undergone great progress thanks to three factors: the development of more sophisticated algorithms, the increase in computing capacity and access to enormous amounts of data. These factors have allowed AI to be applied to areas as diverse as medicine, finance, meteorology or transport, among others.

According to Rouhiainen (Rouhiainen, 2019)In the research entitled, "Artificial Intelligence for Companies", he mentions that Artificial Intelligence is drastically changing the business environment and how companies work. It is the most important technology of our times as it is capable of both breaking with traditional processes and empowering those companies that apply it correctly. It should be noted that in his research he says that if companies invested in AI and in the collaboration between humans and machines, they could increase revenues by 38% and raise employment levels by 10% in 2018 and 2022.

In the opinion of Pajeres et.al (Pajares Mrtinsanz, 2006)they define it as: “Artificial Intelligence is a science oriented to the design and construction of machines that implement tasks typical of humans endowed with intelligence”

From Franganillo's point of view, he mentions that (Franganillo, 2023)”Generative Artificial Intelligence (AI) is a branch of computer science that deals with the automatic creation of content of various kinds, such as texts, images, cloned voices, and videos. To do this, it uses deep learning techniques and models trained on large volumes of data with which it is able to produce original and realistic content that is often indistinguishable from that generated by humans.”

Within the Economy, AI refers to the use, implementation and application of intelligent technologies to improve efficiency, productivity and decision making in various business sectors.

There are several fields where AI is applied.

That is why we will mention some areas that have had an impact on the economic aspect.

Relevant aspects with the use of AI Audience segmentation - Market segmentation (leads).

Economic Growth Identifying opportunities 27/7 communication with customers (chatbot).

AREAS OF AI IN THE ECONOMY

AI in Finance

According to Matthew Finio (Matthew Finio, 2023) AI in finance is the use of technology, including advanced algorithms and machine learning (ML), to analyze data, automate tasks, and improve decision-making in the financial services industry. This financial technology enables financial services organizations to improve the efficiency, accuracy, and speed of tasks such as data analytics, forecasting, investment management, risk management, fraud detection, customer support, and more. AI is modernizing the financial industry by automating traditionally manual banking processes, enabling a better understanding of financial markets and creating ways to engage customers that mimic human intelligence and interaction.

Analyzing consumer trends and behaviors represents another valuable facet of integrating (LLMs) into financial services. These models can analyze large amounts of data to identify market trends, consumer behaviors, and associated sentiment. This capability allows financial institutions to adjust their strategies and offerings more effectively, adapting to changing market dynamics and improving their agility in the face of evolving expectations. (Garmendia, 2024)

This approach also opens up the possibilities to transform the finance function into different use cases to help reduce costs, accelerate processes and increase productivity, for example:

- Automation of reports. Generative AI can simplify both internal and external financial reporting. By linking financial data sources to this technology, users can request the consultation, summary and interpretation of databases, allowing the financial area to analyze and present the information more quickly.
- Scenario generation. GenAI also provides the finance function with a tool for strategic planning. For example, both internal and external data can be used to model scenarios that simulate particular trends, such as in inflation or exchange rates. The results obtained help to make better decisions.
- Data integration for compliance. Artificial intelligence can be used to collect and analyze financial data to identify potential errors in regulatory, auditing, or legal issues. This early detection will allow companies to respond proactively to better manage the risks of non-compliance. (Jiménez, 2024)

One of the things that is helping the financial area is:

- Algorithmic trading: AI can be used to develop trading algorithms that can analyze market trends and historical data to make decisions and execute trades faster than humans.
- Automation and efficiency: AI can automate repetitive and time consuming tasks, allowing financial institutions to process large amounts of data faster and more accurately.
- Competitive advantage: AI can help financial institutions foster innovation and stay ahead of technology, which can give them a competitive advantage.
- Compliance: AI can automate monitoring and reporting requirements to ensure regulatory compliance.

AI in Marketing

According to Bustos (Bustos, 2017) digital marketing is defined as the application of marketing strategies that are carried out through digital media, allowing information to be shared in real time, advertising strategies to be carried out, and thus allowing the identification of the impact on users.

On the other hand, Chunque (Chunque, 2021) states that digital marketing is important since it seeks to improve and optimize the performance and position of companies within the markets, and therefore giving digital presence is a strategy in order to achieve greater visibility before users or consumers Artificial intelligence marketing is a method of leveraging intelligence technologies to collect data, customer insights, anticipate customers' next moves, and make automated decisions that impact marketing efforts.

According to Kotler (Philip Kotler), the most important function of digital marketing is to drive action and encourage consumer word of mouth. Since digital marketing is more easily assessable than traditional marketing, the former focuses on improving results while the objective of the latter lies in initiating interaction with consumers.

AI can boost marketing return on investment (ROI) due to its impact on productivity. Where to use AI in Digital Marketing?

Data Analysis and Audience Segmentation: AI can analyze large amounts of data in real-time and segment the audience into specific groups based on their interests, behaviors, and preferences. This allows businesses to create personalized and targeted messages to each segment, increasing the relevance and effectiveness of campaigns.

Process Automation: AI can automate repetitive and routine tasks, such as sending emails, publishing content on social media, and managing advertising campaigns. This frees up time and resources for marketers to focus on strategic and creative activities.

Chatbots and Customer Support: AI powered chatbots can provide assistance and answer customer questions in real time. These chatbots can simulate human conversations and offer a faster and more effective customer service experience.

Content Personalization: AI allows the personalization of content based on user behavior and preferences. By analyzing customer data, AI can offer recommendations and relevant content for each individual, improving the customer experience and increasing conversion rates.

Ad Campaign Optimization: AI can optimize ad campaigns by analyzing real time data and adjusting bids and messaging to maximize return on investment (ROI). This allows for greater accuracy in audience targeting and ad budget allocation.

Trend Forecasting: AI can analyze historical data and behavioral patterns to forecast future market and consumer behavior trends. This helps businesses be prepared for changes in demand and make informed decisions in their marketing strategies.

AI in Accounting

In Marcela's opinion (Marcela del Carmen Sanchez Balcázar, 2023), artificial intelligence (AI) has also begun to have an impact on the field of accounting. AI can automate repetitive, low complexity tasks, such as invoice processing, bank reconciliations, and financial reporting. This frees up time for accountants to focus on more strategic, higher value added activities

According to (Angarita, 2023), explains that the need to implement AI in public accounting can allow various improvements in the accounting and financial processes carried out by the accountant since the world is currently incorporating different technologies derived from artificial intelligence through various means, such as social networks, and/or platforms.

Some of the benefits implemented in accounting are:

Automation: You can automate repetitive tasks, freeing accountants from repetitive or monotonous workload and allowing them to focus on more creative and strategic activities.

Fraud reduction: You can identify patterns and anomalies in financial data, helping to prevent fraud and ensure greater accuracy in accounting records.

Improved efficiency: You can speed up the process of data entry and analysis by recognizing and categorizing transactions from images, saving time and reducing errors.

Reporting: Automatic financial reporting is a task where AI can excel and will contribute significantly to streamlining the accounting process.

Independent auditing: It can serve as an independent auditing mechanism by evaluating transactions and reports, identifying patterns, and detecting irregularities in financial data. (La IA en la contabilidad: beneficios y riesgos asociados, 2023)

The use of artificial intelligence for decisions in an organization.

Technological advances continue to grow and are increasingly present in people's daily lives, as they are more familiar with them and their use becomes easier. This positive impact is very noticeable when the advances are related to artificial intelligence (AI). The impact of technology and AI has also benefited different areas such as organizations or companies, where many of them have decided to invest in technology to optimize their processes, considering science that includes intelligent models and machines that allow imitating human intelligence, automating decision-making and problem solving.

Organizations such as Walmart and Coca-Cola have tested the use of robots in stores to guide customers to the aisles and shelves where their products are located, thus improving the shopping experience. Alibaba, Amazon, Apple, Meta, Tesla, Netflix, Google and Microsoft are also leading the way towards a digital revolution. Business prospects point to exponential growth in the adoption of AI in the short and medium term, as this technology is reconfiguring processes, ideas and interests among companies. (Ana Franco Valdez, 2024)

Nowadays, technology companies are investing in artificial intelligence, improved algorithms and machine learning. One of the sectors in which e-commerce companies invest the most is in machine learning, in which machines are able to recognize products from a simple image. To know what product it is, it is not necessary to indicate the reference, the brand or any written information, simply the image of the product and speed up and speed up the processes. This has been possible thanks to supervised learning and machine learning techniques where they have been trained with many images of a certain product from different angles so that it is able to recognize it with high precision. (Zorrilla, 2022)

An example is Amazon, which is an e-commerce company. The company uses a large number of artificial intelligence algorithms, of which there are several that are published. Thanks to these algorithms, Amazon has been able to develop. One of the most important algorithms that Amazon uses to apply and optimize its search engines is SEO (Search Engine Optimization). In this way, the consumer will find exactly what they are looking for in a faster way. This algorithm has been renamed A9 and its purpose is to sell as much as possible and is used in most marketplaces. (Zorrilla, 2022)

Currently, since there are a large number of vendors selling on Amazon and it is a very competitive platform, one of the ways to attract the attention of consumers is by optimizing or paying for ads by vendors within the website. With this algorithm, sales and profits increase since the search results for consumers are positioned based on their taste or what they want to buy. This algorithm works by using keywords and classifies the products in the ranking of the page. Customers perform their searches using terms related to the name or description of the product and the program is responsible for classifying them. (Nicolás, 2021)

One of the most impactful and intelligent algorithms that Amazon has is the one it uses when setting prices for its products. Amazon has a system and algorithms whose function is to obtain the lowest prices that competitors have for the same product, so that the company will always have the lowest prices on the market for its customers.

On the other hand, Amazon can predict future purchases or recommendations to customers, for this it has stored all the purchases made previously, as well as the searches made. Each user or client to be able to make a purchase through Amazon has to have created a personal account so that all the data generated is associated with the account. Taking this into account, Amazon compares the purchases that a client has made with those made by others, in case of having similarities in the characteristics it recommends it to have a greater impact and probability of being able to acquire the product (Zorrilla, 2022)

Through the algorithms used by Amazon, large databases (Big Data) have been generated. This Big Data is used to make both short and long-term decisions. Thanks to machine learning and the creation of algorithms, Amazon has created prediction and analysis techniques to improve navigation on its website. This has been carried out thanks to the capacity of the systems to learn through the analysis of massive data obtained from the collection, storage and ordering that Big Data does. (Amazon y el Big Data. Una historia de éxito. (s.f.)., s.f.)

CONCLUSIONS

It is important to mention that AI is present in the different organizational areas, we must consider that the application of these tools favors decision making, in terms of their effectiveness, efficiency, productivity and above all customer satisfaction, attraction of new prospects (leads).

It is worth mentioning that each of the tools are also exposed to information theft, which is why every day companies specialize in cybersecurity in order to reduce risks and protect this information.

REFERENCES

AI in accounting: benefits and associated risks. (2023, November 20). Retrieved from https://incp.org.co/publicaciones/boletin-virtual/2023/11/la-ia-en-la-contabilidad-beneficios-y-riesgos-asociados/

Amazon and Big Data. A success story. (n.d.). Retrieved from https://ayudaleyprotecciondatos.es/big-data/amazon/

Angarita, G. T. (2023). *The impacts of artificial intelligence on the role of the public accountant*. Faculty of Economic, Administrative and Accounting Sciences, Cooperative University.

Boden, M. A. (2018). *Artificial intelligence: A very short introduction*. Oxford University Press. DOI: 10.1093/actrade/9780199602919.001.0001

Bustos, E. V. (2017). *Digital marketing*.

Chunque, V. (2021). *Digital marketing and positioning level of the company*.

Finio, M., & D., A. (2023, December 8). *Artificial intelligence in finance*. IBM. Retrieved from https://www.ibm.com/mx-es/topics/artificial-intelligence-finance

Franco Valdez, A. (2024). *The implementation of artificial intelligence in retail commerce*. Retrieved from https://centrocomerciodetallista.tec.mx/sites/g/files/vgjovo1571/files/CCD_Inteigencia%20Artificial%2003

Franganillo. (2023). *The large language models. ThinKEPI Yearbook, 17*.

Garmendia, L. S. (2024, March 15). *Beyond the numbers: This is the AI that will take the financial sector into the future*. Retrieved from https://forbes.com.mx/mas-alla-de-los-numeros-asi-es-la-ia-que-llevara-al-sector-financiero-al-futuro/

Jiménez, A. P. (2024, April 30). *Artificial intelligence in the financial sector*. Retrieved from https://www.pwc.com/mx/es/liderazgo-estrategico/inteligencia-artificial-en-el-sector-financiero.html

Kotler, P., & Kartajaya, H. (n.d.). *Marketing 4.0*. LID.

Kurian, N., James, D., Varghese, V. S., Cherian, J. M., & Varghese, K. G. (2023). Artificial intelligence in scientific publications. *The Journal of the American Dental Association, 154*(12), 1041–1043. DOI: 10.1016/j.adaj.2023.03.015 PMID: 37140497

Nicolás, J. J. (2021, May 18). *Getting to know A9: Dissecting Amazon's algorithm.* Retrieved from https://es.linkedin.com/pulse/conociendo-a9-destripando-el-algoritmode-amazon-jose-javier-nicol%C3%A1s

Pajares Mrtinsanz, G., & García, S. (2006). *Artificial Intelligence and Knowledge Engineering*. Alfaomega.

Rouhiainen, J. (2019). *Artificial intelligence for business.* Retrieved June 11, 2024, from https://repositorio.usil.edu.pe/server/api/core/bitstreams/e190d249-c1ca-4006-bf5f-cf1b0d0a12cd/content

Sánchez Balcázar, M. C., & Gómez, E. (2023). *Effects of the application of artificial intelligence in accounting and decision-making*. https://revistap.ejeutap.edu.co/index.php/Gestion/article/view/71

Zorrilla, R. A. (2022). *Ethics and the use of algorithms and artificial intelligence by Amazon: A value analysis.* Retrieved from https://repositorio.comillas.edu/rest/bitstreams/527617/retrieve

Chapter 10
Sustainable Cities: A Pathway to Achieving UN Sustainable Development Goal 11

Gargee Gavhane
Ajeenkya D.Y. Patil University, India

Atharva Saraf
https://orcid.org/0009-0006-3842-5508
Ajeenkya D.Y. Patil University, India

Alishba Siddiqui
Ajeenkya D.Y. Patil University, India

Susanta Das
https://orcid.org/0000-0002-9314-3988
Ajeenkya D.Y. Patil University, India

Rebika Singh
Ajeenkya D.Y. Patil University, India

Kamaljeet Kaur Siddhu
Ajeenkya D.Y. Patil University, India

Nishakant Ojha
Ajeenkya D.Y. Patil University, India

Rakesh Kumar Jain
https://orcid.org/0000-0002-0216-5437
Ajeenkya D.Y. Patil University, India

ABSTRACT

This chapter presents a comprehensive analysis of UNSDG 11, i.e., making

DOI: 10.4018/979-8-3693-8643-9.ch010

human settlements and cities inclusive, safe, resilient, and sustainable. As the world is facing increased urbanization, with an estimated 70% of the global population projected to reside in cities by 2050, the demand for sustainable urban development has never been so urgent. On the basis of an analysis of theoretical models, empirical evidence, policy interventions, and case studies from diverse global contexts, this chapter outlines the multifaceted nature of urban sustainability challenges and opportunities. This chapter also discusses the convergence of technological advancements, including artificial intelligence and smart city technologies, in advancing urban sustainability goals. Through the convergence of knowledge from urban planning, environmental science, sociology, economics, and technology studies, this chapter develops in-depth understanding of opportunities for the fulfillment of SDG 11 under the pressure of increased urbanization, climate change, and socioeconomic inequality.

1. INTRODUCTION

The United Nations' Sustainable Development Goal 11 (SDG 11) calls for making cities and human settlements inclusive, safe, resilient, and sustainable. It is predicted that 70% of the whole world's population will be living in cities by 2050 (United Nations Department of Economic and Social Affairs [UN DESA], 2023), so it is more important than ever that the complex issues of urbanization are tackled. This chapter examines SDG 11 in depth, referring to important areas like urban poverty, public transport accessibility, air quality, infrastructure, public spaces, disaster risk reduction, public awareness, and the transformative potential of artificial intelligence (AI). This chapter gives an in-depth analysis of the existing strategies, defines existing gaps, and suggests workable solutions for promoting sustainable urban development in an ever-changing, complex, and multidimensional urban world.

Urban sustainability is one of the hallmark challenges of the 21st century. Cities cover just 3% of the land area but consume 60-80% of energy and 75% of carbon emissions (UN-Habitat, 2023). Beyond their environmental impact, urban settlements continue to be engines of economic growth, centers of innovation, and hotspots of cultural exchange and social progress. The Janus-like character of urban settlements—sources of problem and solution at the same time—resonates with the maturity of realizing SDG 11.

Sustainable urbanity transcends traditional green issues to encompass social justice, economic viability, and institutional capacity. Sachs et al. (2021) argue that urban sustainability requires an integrated approach that balances environmental protection with social equity and economic development. This is articulated in the ten SDG 11 targets, which collectively address provision of housing, transport, participatory urban planning, cultural heritage, disaster risk, environmental footprint reduction, provision of public space, and rural-urban linkages (United Nations, 2024). This chapter aims to provide a critical reflection on SDG 11, in theory and practice, in a range of different urban settings around the world. Critically evaluating current achievement, current challenges and new potential, it aims to contribute to current debate in sustainable urban development, and to set the agenda for future policy and practice.

2. CONCEPTUAL MODELS OF SUSTAINABLE URBAN DEVELOPMENT

2.1 Urban Sustainability Conceptualization

The theoretical underpinnings of sustainable urbanization have evolved significantly over the past decades, as shifting perceptions of the relationship between urbanization and environmental impacts have arisen. The early articulations of urban sustainability were environmentally in focus, with particular attention to minimizing the use of resources and pollution (Beatley, 2012). Later articulations, however, have incorporated social equity, economic sustainability, and governance dimensions.

Campbell's (1996) "planner's triangle" model names three basic priorities of sustainable development as environmental protection, economic growth, and social equity. The model shows tensions between the priorities that are necessary and encourages proportionate reactions to urban development and planning. Connelly (2007) elaborates on the theory by saying that sustainable urban development means resolving the "conflicts in definition" that occur when stakeholders define sustainability in a way that makes it serve their interests and values.

Most recently, Raworth's (2017) "doughnut economics" framework has been used more and more as a framework for describing urban sustainability. The framework imagines a "safe and just space for humanity" between a social base of wellbeing and an ecological limit of planetary boundaries. Amsterdam and Portland cities have begun implementing this framework already to inform their sustainability policy (Amsterdam City, 2023).

2.2 Systems Thinking and Urban Metabolism

The concept of urban metabolism, in which cities are viewed as advanced systems with energy, material, water, and waste flows, is increasingly being accepted in sustainable urban planning (Kennedy et al., 2011). The approach enables detailed examination of resource use patterns and waste production and can be employed to inform actions to improve urban resource efficiency.

Wolman's (1965) pioneering work on urban metabolism set the stage for imagining cities as systems with inputs, outputs, and internal workings. Contemporary scholars have expanded this model to include social aspects, considering how resources flow and interact with issues of equity, justice, and wellbeing (Ramaswami et al., 2017).

Systems thinking models recognize the interconnection of urban issues and the likelihood of cascading effects between industries. As Bai et al. (2016) explain, fixing sustainability in one sector (say, transport) necessarily has effects in other sectors (such as air, health, economy). This reality underlines the imperative for integrative policy approaches that can fix a set of sustainability problems concurrently.

2.3 Resilience Theory and Urban Sustainability

Urban resilience—the capacity of urban systems to survive, adapt, and recover from shocks and stresses—has become a seminal theme in the literature on sustainable urban development (Meerow et al., 2016). The Rockefeller Foundation's 100 Resilient Cities initiative brought the concept mainstream, and urban resilience is "the capacity of individuals, communities, institutions, businesses, and systems within a city to survive, adapt, and grow no matter what kinds of chronic stresses and acute shocks they experience" (Resilient Cities Network, 2023).

Resilience thinking emphasizes the dynamic nature of urban systems and their capacity to transform and adapt. Folke (2006) makes a distinction between engineering resilience (ability to recover to a steady-state after disturbance) and ecological resilience (ability to absorb disturbance and yet maintain function). Transformative resilience—the ability to reorganize systems in a fundamental way in response to changing conditions—is particularly relevant in urban sustainability under the conditions of climate change and other chronic stressors (Elmqvist et al., 2019). The intersectional theory of sustainability and resilience has yielded productive tensions and learning. Resilience has been viewed by some scholars as a prerequisite for sustainability (Marchese et al., 2018), while others caution that resilient systems can perpetuate unsustainable modes of functioning unless guided by normative sustainability principles (Olsson et al., 2015). This debate calls for the development of integrated frameworks that leverage the analytical potential of resilience thinking without sacrificing the normative aim of sustainability.

3.THE ROLE OF SUSTAINABLE CITIES

3.1 Urbanization Trends and Challenges

The speed and scope of universal urbanization present challenges and opportunities for sustainable development. The number of people living in urban areas is likely to increase by 2.5 billion to 2050, with nearly 90% of the growth occurring in Asia and Africa (UN DESA, 2023). This will demand unprecedented investment in urban infrastructure, housing, and services.

Urbanization presents vast economic opportunities, innovation, and improved living standards. However, it is accompanied by massive challenges such as scarce housing, pollution, inefficient transport systems, and vulnerability to climate change. SDG 11 aims at making cities have environments where all people can live in healthy and safe conditions with access to affordable and sustainable shelter, clean air, efficient transport, and climate-resilient infrastructure.

Urbanization is extremely varied by country and region. Urbanization in the developed world has been incremental, with stepwise institution and infrastructure adjustment, while in the majority of developing countries, rapid and often uncontrolled urbanization is occurring (Cobbinah et al., 2015). Such a "compressed urbanization" is particularly challenging for sustainable development, since the provision of infrastructure trails population increase (Chen et al., 2014).

Mega-cities—clusters of cities with over 10 million residents—are a feature of contemporary urbanization. Mega-cities across the globe increased from 10 in 1990 to 33 in 2023 and are projected to increase to 43 mega-cities by 2030 (UN DESA, 2023). Mega-cities are also characterized by sustainability concerns that are focused on size, sophistication, and their impact on their respective regions.

3.2 Economic Dimensions of Urban Sustainability

Cities generate around 80% of global GDP, and they are the sources of economic growth and innovation (World Bank, 2022). Economic concentration in cities generates agglomeration economies—efficiency gains resulting from geographic concentration of consumers, workers, and firms (Glaeser, 2011). Such efficiency gains result in greater compensation, greater innovation, and greater economic opportunity.

Still, economic gains from urbanization are unequally distributed, with colossal differences between formal and informal sectors, central and peripheral areas, and various socioeconomic groups. As Sassen (2018) writes, modern urbanization processes are bound to produce "expulsions"—the deliberate exclusion of marginalized groups from economic growth.

Inclusive economic policies must address such inequalities for sustainable city development. The "just city" approach, which Fainstein (2010) champions, highlights the need for equity, democracy, and diversity in economic development at the city level. Likewise, the "right to the city" approach, which is an adaptation of Lefebvre's (1968) thought, presumes the right of all city dwellers to access and enjoy urban processes.

The COVID-19 crisis exposed both the economic vulnerability and potential resilience of cities. While urban economies were hit hard, particularly in industries that depend on physical proximity and face-to-face interaction, the majority of cities demonstrated high resilience through remote working practices, digital delivery of services, and local supply chain innovation (OECD, 2023a). These experiences emphasize the importance of economic diversity and adaptability as an urban sustainability factor.

3.3 Environmental Imperatives for Sustainable Urban Development

Urbanization is deeply environmentally impactful at the local, regional, and global levels. Urban processes cause cities to be significant drivers of climate change, such that cities produce 70% of global greenhouse gas emissions while occupying only 2% of the Earth's surface (C40 Cities, 2023). Without radical transformation of the form, character, and governance of cities, projected urbanization can entrap carbon-intensive growth for decades to come.

Besides climate change, cities also experience numerous environmental issues, including air and water pollution, waste, loss of biodiversity, and heat island effect. The World Health Organization estimates that 90% of urban dwellers are breathing air that falls below WHO standards, resulting in 4.2 million premature deaths annually (WHO, 2023). Similarly, inefficient waste management in most cities is resulting in environmental degradation, public health risks, and greenhouse gas emissions from landfills.

Environmental concerns disproportionately affect marginalized groups, hence creating environmental justice concerns. The poor are exposed to more pollution, have less green space, and are more exposed to climate change (Agyeman et al., 2016). These disparities must be addressed through the incorporation of environmental sustainability and social equity in decision-making.

Despite all these challenges, the city also provides unique opportunities for environmental sustainability. Compact urbanization can potentially reduce per capita energy consumption and carbon dioxide emissions through optimized infrastructure, shared facilities, and reduced transport needs (Newman & Kenworthy, 2021). Copenhagen, Stockholm, and Portland have demonstrated the potential for absolute decoupling of economic growth and environmental pressures through integrated sustainability policies (Global Covenant of Mayors, 2023).

3.4 Social Dimensions of Urban Sustainability

Social sustainability—equity, inclusivity, wellbeing, and quality of life—is a core component of sustainable urban development. Dempsey et al. (2011) identify a range of the most important factors of urban social sustainability as social interaction, collective group participation, stability in communities, pride/sense of place, security, and access to opportunity.

Urban inequality is complex and impacts multiple dimensions including income, housing, education, health, and access to basic services. UN-Habitat (2023) approximates that nearly one billion individuals live in slums or informal settlements with low living conditions, poor water and sanitation, and insecure land tenure. Spatial manifestations of inequality undermine social cohesion and constrain human development potential.

Soja's (2010) concept of "spatial justice" speaks to the way urban space re-produces social injustice. This school of thought is interested in the way the geography of opportunity—the spatial distribution of resources, services, and amenities—impacts life chances and social outcomes. Urban planning interventions, housing policy interventions, and public service provision must confront spatial injustice in order to equitably provide access to urban resources. Social sustainability also entails cultural features like preservation of cultural heritage, improvement of cultural diversity, and public spaces inclusive of all. As UNESCO (2022) points out, culture is a "transversal and fundamental component of sustainable development" that assists in improving social cohesion, economic opportunity, and environmental conservation. Those cities that value and nurture their cultural capital can foster quality of life as well as resilience capacities.

4. KEY COMPONENTS OF SUSTAINABLE CITIES

4.1 Affordable and Adequate Housing

Housing is a basic human right and need, the cornerstone of health, wellbeing, economic engagement, and social inclusion. However, housing affordability and quality are long-standing issues in cities globally. The UN estimates that more than 1 billion people globally reside in slums with poor housing and infrastructure (UN-Habitat, 2023). The housing crisis has several dimensions, such as affordability limitations, quality deficiencies, location disadvantage, and tenure insecurity.

Housing affordability has worsened in most cities because of various factors, such as speculation, land-use constraints, inflation in construction costs, and stagnant wages. Based on OECD statistics (2023b), more than 40% of low-income households' disposable incomes go toward the cost of housing in most large cities, essentially locking the households in a financially precarious cycle.

Policies like rent control, affordable housing programs, and slum upgrading programs are essential to meet these challenges. Inclusionary zoning policies, which mandate or encourage developers to provide affordable units in market-rate projects, have been effective in cities like Vienna, Singapore, and Boston (Calavita & Mallach, 2010). Likewise, community land trusts—nonprofit organizations that hold permanent title to land and sell or lease homes at below-market prices—have proven effective in producing permanently affordable housing in cities like Burlington, Vermont and Bristol, England (Davis, 2010). Brazil's Minha Casa Minha Vida (My House, My Life) program is one of the world's largest affordable housing programs, having produced around 4.3 million housing units since its launch in 2009 (Rolnik et al., 2021).

Although the program has greatly increased access to housing, assessments have noted issues of peripheral location, restricted access to urban services, and social segregation (Cardoso & Jaenisch, 2017). These results underscore the need to tie housing provision together with more comprehensive urban planning goals to provide for access to jobs, schools, health care, and transportation. Innovative strategies for housing delivery involve adaptive reuse of existing structures, incremental housing types that provide room for gradual expansion and upgrading, and co-housing types that integrate private homes with common facilities. These strategies can promote both affordability and sustainability while creating social connections and community resilience (Tummers, 2016).

4.2 Sustainable Transport and Mobility

Transportation systems significantly influence urban form, function, and quality of life. Effective, accessible, and sustainable urban mobility supports economic opportunity, social inclusion, and environmental sustainability. In contrast, inadequately planned transportation systems can increase congestion, pollution, inequality, and carbon emissions.

The prevalence of private cars in most urban transport systems has produced many sustainability issues such as traffic congestion, air pollution, greenhouse gas emissions, and land use inefficiency. Economic losses due to congestion alone amount to 2-5% of GDP in most major cities (Stickler, 2022). In addition, auto-oriented urbanization patterns tend to exclude non-drivers, such as low-income residents, seniors, children, and people with disabilities.

Sustainable transport in cities needs a move towards more sustainable, equitable, and efficient mobility solutions. The "avoid-shift-improve" approach is a helpful framework for transport sustainability policy (Dalkmann & Brannigan, 2007):

> Avoid: Minimize the necessity for motorized travel by means of integrated land use and transport planning, teleworking, and electronic delivery of services.
> Shift: Encourage modal change from private car use to more sustainable transport modes like public transport, cycling, and walking.
> Improve: Improve the energy efficiency and lower the environmental footprint of all modes through technology and operational advances.

Curitiba, Bogotá, and Singapore, among other cities, have already established Bus Rapid Transit (BRT) systems that offer high-capacity, high-frequency service at lower expense than rail-based options.

The TransMillenial system in Bogotá, for instance, carries around 2.4 million passengers per day and has lowered carbon emissions by some 350,000 tons each year (C40 Cities, 2023).

Likewise, Singapore's comprehensive transport system integrates mass rapid transit, buses, and demand management strategies like congestion pricing and vehicle quota schemes to attain high mobility levels with relatively low environmental effects. Active transport—cycling and walking—is a vital part of sustainable urban mobility. The health effects of active transport are significant, with regular exercise lowering the risk of heart disease, stroke, type 2 diabetes, and some cancers (Mueller et al., 2015). In addition, pedestrian and bicycle infrastructure is often less expensive and requires less space than automobile infrastructure, enabling greater use of urban land and public funds. Electric mobility and cycling-infrastructure encourage environmentally friendly means of transport. Copenhagen and Amsterdam have accomplished about 35% and 27% cycling mode shares, respectively, by means of well-connected networks of dedicated cycling facilities, traffic calming, and favorable policies (European Cycling Federation, 2023). These cities prove that cycling is a mainstream mode of transport, not just a leisure activity, if it is facilitated by suitable infrastructure and policies.

4.3 Air Quality and Pollution Management

Air pollution is one of the most critical environmental health issues in the world, with particularly high effects in urban environments. The World Health Organization estimates that 7 million deaths are caused by air pollution each year, and thus, urban air quality is a top priority for sustainable development (WHO, 2023). Primary causes of urban air pollution are transportation, industrial processes, energy generation, construction, and solid waste burning.

The health effects of air pollution are respiratory illness, cardiovascular disease, stroke, and lung cancer. Children, the elderly, and those with existing medical conditions are at increased risk to these effects. In addition to health consequences, air pollution imposes significant economic burdens in terms of lost productivity, healthcare costs, and decreased agricultural production (Landrigan et al., 2018).

Urban air quality management demands integrated strategies targeting multiple pollutants and sources. Strategies include:

Emission regulations and standards: Impose and apply caps on automobile, industrial, and other emission sources.
Clean energy transition: Transition from fossil fuels to renewables for electricity supply, heating, and transportation.
Sustainable city planning: Encourage dense, mixed-use construction that minimizes transport emissions and promotes natural ventilation.
Green infrastructure: Increasing Forest cover, greenspace, and vegetation in the city to capture pollutants and cool urban areas.
Public perception and behavior modification: Raising awareness among residents on air quality effects and encouraging green practices.

Installing low-emission zones, green belts, and air pollution control policies can help tremendously in enhancing the quality of the air. The Ultra Low Emission Zone in London, installed in 2019 and widened in 2021, decreased nitrogen dioxide levels by 44% in the central area of London (Greater London Authority, 2023). The policy makes it necessary for vehicles coming into the area to adhere to certain emission limits or pay a daily fee, thereby hastening the switch to cleaner vehicles.

4.4 Urban Resilience and Disaster Risk Reduction

As the number of climate-related disasters as well as other shocks grows, cities need to place topmost priority on resilience—the ability to prepare for, respond to, and recover from disruptive events in the face of changing circumstances. Urban areas are exposed to numerous natural hazards, such as floods, storms, heatwaves, droughts, earthquakes, and landslides, along with technological and social threats like infrastructure collapse, pandemics, and civil disorder.

Climate change exacerbates many of these threats through increased and more intense extreme weather events, sea-level rise, and changes in temperature and precipitation patterns. The Intergovernmental Panel on Climate Change (IPCC, 2022) reports that cities are particularly exposed to climate change effects, such as higher mortality and morbidity during heatwaves, increased flood risk, damage to infrastructure, and disruption of essential services. These effects fall disproportionately on vulnerable groups, such as low-income communities, residents of informal settlements, and marginalized groups.

Policies like floodplain zoning, early warning, and resilient infrastructure are some of the most effective strategies for increasing urban resilience. The Sendai Framework for Disaster Risk Reduction 2015-2030 is based on four priorities: understanding disaster risk, strengthening governance structures, investing in resilience, and improving preparedness for effective response and recovery (UNDRR, 2015).

Japan's earthquake-resistant building codes have reduced disaster effects even in a highly seismically active country. After the 1995 Kobe earthquake, which caused about 6,400 fatalities and widespread structural damage, Japan made its building codes and enforcement measures stricter (Kim & Kaneda, 2018). The codes demand that buildings resist high ground motion without collapse to save lives and support post-disaster recovery.

Community-based disaster risk reduction (CBDRR) represents a crucial component of urban resilience, particularly in contexts with limited institutional capacity or trust. CBDRR approaches engage local communities in hazard identification, vulnerability assessment, capacity building, and emergency response planning. Programs such as the Philippines' Barangay Disaster Risk Reduction and Management Committees have demonstrated effectiveness in reducing disaster impacts through local knowledge integration, social capital development, and participatory governance (Fernandez et al., 2019).

4.5 Smart Cities and Technological Innovations

Technological advances, especially those in artificial intelligence, Internet of Things (IoT), big data analytics, and digital connection, present great opportunities to heighten urban sustainability. The technological applications under "smart city" include these themes, with greater focus on making use of digital technologies to create more efficient and better urban administration, service, and living conditions.

IoT and AI-enabled solutions improve city efficiency, consuming less energy while enhancing service provision. The two technologies allow monitoring and optimization of urban systems in real-time, predictive maintenance of the infrastructure, and customized delivery of services on the basis of specific needs and preferences. An illustration is provided by Barcelona's intelligent traffic system, which deploys sensors, cameras, and algorithms to rationalize traffic flows, lowering congestion and related emissions by about 21% (Barcelona Digital City, 2023).

Smart energy networks combine renewable generation, energy storage, efficient networks, and demand response to cut carbon emissions and improve reliability and affordability. These goals are supported by technologies including smart meters, building automation systems, and grid management software. Copenhagen's smart energy system, for example, integrates district heating, wind energy, and waste-to-energy plants with digital monitoring and control technology to deliver high efficiency and utilization of renewable energy (State of Green, 2023).

Singapore's AI-based waste management is the best example of how smart technologies can solve urban environmental issues. The system utilizes sensors to track waste in bins, plan optimal collection routes, and analyze the composition of waste for better recycling. These uses have lowered collection expenses by around 20% while cutting vehicle emissions and landfill disposal (Singapore Government Technology Agency, 2023).

5. CHALLENGES IN ATTAINING SDG 11

5.1 Economic and Financial Limitations

Urban infrastructure development involves high investments, but most developing countries lack funding. The World Bank (2023) estimates that it will cost around $4.5-5.4 trillion per year to attain SDG 11 globally, with a wide financing gap in low- and middle-income countries. This gap captures several reasons such as the inability of municipalities to generate revenues, limited access to capital markets, and alternative uses of available resources.

Public-private partnerships (PPPs) have proven to be a plausible solution for urban project financing, allowing for public-private risk-sharing and the application of private know-how and finance. Good practice is represented by Sao Paulo's Line 4 metro project, where public financing was combined with private operation and maintenance, as well as Manila's water service concessions, which increased coverage at the same time as efficiency improved (World Bank, 2023).

Innovative financing instruments like green bonds, social impact bonds, land value capture, and climate finance offer other avenues through which to finance sustainable urban growth. Green bonds—specialized debt instruments for environmental initiatives—have expanded quickly, with yearly issuance increasing from $37 billion in 2014 to more than $500 billion in 2023 (Climate Bonds Initiative, 2023).

5.2 Governance and Policy Gaps

Inadequate urban planning, the absence of coordination among government agencies, and an ineffective policy implementation hold back efforts towards SDG 11. Urban governance refers to the institutions, processes, and relations through which decisions concerning urban development are made and carried out. Effective governance needs suitable legal systems, institutional setups, financial means, and participatory mechanisms (UN-Habitat, 2020).

Most cities, especially in quickly urbanizing areas, have challenged governance systems that are fragmented, with ill-defined mandates, multiple overlapping jurisdictions, and little coordination. The problems of fragmentation most commonly arise through isolated sectoral policies to respond to transportation, housing, the environment, and economic development as separate components, not as parts of integrated urban systems (OECD, 2023c).

Decentralized city governance models, such as in Sweden and Switzerland, have proved effective in simplifying city planning and implementation. Such models delegate significant power and resources to city governments, facilitating context-specific policies and citizen participation. Stockholm's district heating system, managed by a municipal company with great autonomy, illustrates the possibilities of improved environmental sustainability from decentralized governance for the provision of green infrastructure (Rutherford, 2014).

5.3 Environmental Sustainability Challenges

Increasing greenhouse gas emissions and improper waste management are among the factors that lead to environmental degradation in cities across the globe. Cities are responsible for about 70% of the world's carbon emissions, mainly from buildings, transportation, and industry (C40 Cities, 2023). Unless there are strong policy measures and technological shifts, urban carbon footprints will continue to grow as cities increase in population and economic output.

Circular economy designs and green urban planning provide solutions for solving such challenges. The circular economy approach prioritizes waste avoidance, material reuse, recycling, and recovery to reduce resource extraction and waste disposal. Urban examples include reuse of building materials, composting food wastes, product sharing platforms, and industrial symbiosis networks (Ellen MacArthur Foundation, 2023).

Green urban planning combines natural components and ecological processes into the built landscape to increase resilience and sustainability. Some of the major elements include constructed wetlands, rain gardens, permeable surfaces, green roofs and walls, and urban forests. They offer a variety of environmental benefits such as air quality improvement, carbon sequestration, urban heat island reduction, stormwater management, and biodiversity support (Pauleit et al., 2019).

6. CASE STUDIES: BEST PRACTICES IN SUSTAINABLE URBAN DEVELOPMENT

6.1 Copenhagen, Denmark: A Carbon-Neutral City

Copenhagen has set a goal to be the world's first carbon-neutral city by 2025, having already cut carbon emissions by 70% since 1990 (Copenhagen Municipality, 2023). This ambitious target is indicative of the city's long history of commitment to environmental sustainability, social justice, and quality of life.

The Copenhagen Climate Plan comes with several strategies, such as renewable energy transformation, green mobility, energy-efficient construction, and waste management enhancement. The city boasts one of the largest district heating systems globally, providing around 98% of dwellings with heat sourced mainly from renewable biomass and waste incineration (State of Green, 2023).

Copenhagen's transport network is focused on sustainable mobility modes, with cycling responsible for about 35% of all journeys to work or study. Cycling infrastructure in the city is well developed, consisting of 382 kilometers of separated cycle tracks, cycle superhighways between suburbs and the city center, and bicycle-friendly traffic lights (European Cycling Federation, 2023).

6.2 Medellín, Colombia: Socially Inclusive Urban Transformation

Once known for crime and inequality, Medellín revitalized itself through urban mobility projects and social integration. The city's transformation began in the early 2000s with a focus on "social urbanism"—an approach that prioritizes public infrastructure investments in marginalized neighborhoods to reduce inequality and enhance quality of life (Wirth et al., 2022).

The urban cable car network (Metro cable) is a flagship program, linking long-isolated hillside neighborhoods to the metro system and city-center economic opportunities. This intervention shortened travel times from more than two hours to less than half an hour for many residents, promoting access to work and social inclusion (Dávila, 2023).

Complementary investments such as library parks, good schools, business incubators, and public spaces in poor neighborhoods. These interventions have led to notable decreases in violence, with homicide rates falling more than 90% from the early 1990s to 2023 (Medellín Municipality, 2023).

6.3 Singapore: A Smart and Sustainable Urban Model

Singapore incorporates parks, water management, and intelligent technology into its urban planning, with a high degree of sustainability in spite of limits on available space and natural resources. The city-state's holistic approach involves designing housing, transportation, energy, water, waste management, and digital government (Singapore Urban Redevelopment Authority, 2023).

The Smart Nation Initiative by the government maximizes transport, housing, and environmental sustainability through digital technologies. Some of the applications are predictive maintenance of infrastructure, real-time monitoring of environmental conditions, smart waste collection systems, and smart traffic management (Smart Nation Singapore, 2023).

Water management is one of the key strengths of Singapore's sustainability strategy. The "Four National Taps" policy integrates water importation, rainwater collection, reclamation of used water (NEWater), and desalination to achieve water security. These initiatives have turned water vulnerability into resilience in the face of population increase and economic growth (PUB Singapore, 2023).

7. THE ROLE OF STAKEHOLDERS IN PROMOTING SDG 11

7.1 National and Local Governments

Governments are at the forefront of policymaking, urban planning, and investments in infrastructure for sustainable cities. National governments set the legal context, decide on resource allocation, and coordinate action across sectors and jurisdictions. Local governments bring context-specific expertise, implementation capability, and direct contact with citizens (UN-Habitat, 2020).

China's Sponge Cities program is a prime example of national-local government partnership for urban sustainability. Initiated in 2015, the initiative seeks to manage urban flooding, water shortages, and pollution by using nature-based solutions like permeable pavements, rain gardens, and constructed wetlands. The national government offers technical support and funding, while local governments tailor solutions to individual urban conditions (Li et al., 2020).

Efficient governance for sustainable urbanization needs vertical (between governance levels) and horizontal (between policy sectors) integration. Multilevel governance setups like metropolitan planning organizations, regional development agencies, and urban-rural arrangements can promote coordination with respect to subsidiarity principles (OECD, 2023c).

7.2 Private Sector and Corporate Responsibility

Firms that invest in green infrastructure, renewable energy, and sustainable buildings make a significant contribution to SDG 11. The private sector offers finance, knowledge, and innovation capacity that is crucial for urban transition towards sustainability. Corporate engagement varies from creating sustainable products and services to adopting sustainable business practices and participating in community development programs (World Business Council for Sustainable Development, 2023).

Tesla's urban battery storage projects are a paradigm of private sector investment in sustainable infrastructure. The company's Powerpack systems allow buildings, communities, and utilities to store clean energy, minimize peak demand, and make grids more resilient. Initiatives like the 100-megawatt Hornsdale Power Reserve in South Australia showcase the promise of battery storage in enabling renewable energy integration while ensuring greater electricity reliability and affordability (Tesla, 2023).

In addition to discrete sustainability initiatives, companies increasingly appreciate the value of integrated environmental, social, and governance (ESG) strategies. Unilever, Microsoft, and Siemens are among the companies that have set ambitious sustainability goals, such as carbon neutrality pledges, circular economy strategies, and sustainable supply chain management (Corporate Knights, 2023).

7.3 Civil Society and Community Engagement

Urban projects initiated by the community increase inclusivity and sustainability by integrating local knowledge, shared ownership, and building social capital. Civil society organizations enable citizen engagement, represent marginalized communities, and ensure public and private sectors' accountability for sustainability promises (Castán Broto & Westman, 2020).

Porto Alegre's participatory budgeting in Brazil empowers citizens to determine urban investment priorities through a formal deliberative process. The method, initiated in 1989, facilitates citizens to set priorities for infrastructure projects, public services, and community development projects. The democratic innovation has been replicated in thousands of municipalities globally, showing the potential of citizen participation to increase resource allocation efficiency, government transparency, and social inclusion (Cabannes, 2015).

Urban sustainability is supported by community-led climate action efforts, including Transition Towns, which offer a second channel of civil society engagement in urban sustainability. These efforts harness local networks, resources, and knowledge to establish community-owned renewable energy projects, local food systems, repair cafes, and other sustainability measures (Hopkins, 2019). Grassroots efforts support mainstream governance processes while fostering social cohesion and adaptive capacity.

8. FUTURE DIRECTIONS: THE PATH BEYOND 2030

Although the target year of SDG 11 is 2030, sustainable city construction efforts should keep going beyond that time. Priority areas in the future are:

Green bonds and SDG-aligned investments will be instrumental in funding sustainable urban infrastructure. The market for green bonds has increased dramatically, from $37 billion in 2014 to more than $500 billion in 2023, reflecting rising investor appetite for environmental and social impacts (Climate Bonds Initiative, 2023).

Emerging growth of sustainability-linked bonds, blue bonds (targeting ocean conservation), and transition bonds (facilitating transitions away from carbon-heavy and toward low-carbon activities) will continue to diversify the sustainable finance universe.

Blended finance strategies—leveraging public, private, and philanthropic capital—hold specific potential for sustainability issues in low- and middle-income countries. Through the utilization of public capital to de-risk private investment, blended finance is able to deploy capital for initiatives with high social and environmental payoffs but perhaps lower or more delayed financial yields (Convergence, 2023).

Policies for climate change mitigation will increasingly be necessary as global warming accelerates. Urban adaptation to climate necessitates holistic strategies covering the resilience of infrastructure, ecosystem services, reduction of social vulnerability, and capacity development in governance (IPCC, 2022).

Nature-based solutions that utilize ecosystem processes to tackle urban problems—will most likely have increased roles in climate adaptation. Some of these include mangrove restoration to protect coastlines, urban forest growth to moderate temperatures, and constructed wetlands to manage flooding.

9. CONCLUSION

Sustainable cities are the cornerstone of global sustainability, with SDG 11 serving as a crucial guiding framework. This chapter has explored the key components, challenges, and solutions for sustainable urban development, emphasizing the need for collaborative governance, technological integration, and innovative financing models. The transition towards inclusive, resilient, and environmentally conscious cities requires sustained efforts from governments, businesses, and communities alike. By prioritizing sustainability in urban development, cities can enhance the quality of life for billions while safeguarding the planet's future.

REFERENCES

C40 Cities. (2023). Cities are responsible for 70% of global greenhouse gas emissions. https://www.c40.org/researches/cities-are-responsible-for-70-of-global-greenhouse-gas-emissions

C40 Cities. (2023). The impact of Bus Rapid Transit systems in major global cities. https://www.c40.org/initiatives/bus-rapid-transit

C40 Cities. (2023). Cities and climate change: A global perspective. https://www.c40.org

C40 Cities. (2023). How cities are leading the way in climate action. https://www.c40.org/

Agyeman, J., Bullard, R. D., & Evans, B. (2016). *Just sustainabilities: Development in an unequal world.* MIT Press.

Bai, X., Roberts, B., & Chen, J. (2016). Urban sustainability experiments in Asia: Patterns and pathways. *Environmental Science & Policy*, *62*, 1–12. DOI: 10.1016/j.envsci.2016.01.008

Barcelona Digital City. (2023). Smart city traffic management in Barcelona: A case study. https://www.barcelona.cat/digitalcity

Beatley, T. (2012). *Green urbanism: Learning from European cities.* Island Press.

Cabannes, Y. (2015). Participatory budgeting: A powerful, yet underutilized, local governance tool for municipalities. UN-Habitat. https://unhabitat.org/participatory-budgeting-a-powerful-yet-underutilized-local-governance-tool-for-municipalities

Calavita, N., & Mallach, A. (2010). *Inclusionary housing in international perspective: Affordable housing, social inclusion, and land use regulation.* Policy Press.

Campbell, S. (1996). Green cities, growing cities, just cities?: Urban planning and the contradictions of sustainable development. *Journal of the American Planning Association*, *62*(3), 296–312. DOI: 10.1080/01944369608975696

Cardoso, A., & Jaenisch, P. (2017). Challenges in Brazil's affordable housing program: Urbanity, location, and social segregation. *Urban Studies (Edinburgh, Scotland)*, *54*(10), 2289–2308. DOI: 10.1177/0042098016680919

Castán Broto, V., & Westman, L. (2020). Just transitions and the co-production of urban climate solutions. *Urban Studies (Edinburgh, Scotland)*, *57*(7), 1307–1323. DOI: 10.1177/0042098019880670

Chen, Y., Zhang, L., & Li, X. (2014). Compressed urbanization and its impact on sustainable development in China. *Sustainable Cities and Society*, *13*, 1–8. DOI: 10.1016/j.scs.2014.03.003

City of Amsterdam. (2023). *The Amsterdam City Doughnut: A tool for transformative action*. Circle Economy.

Climate Bonds Initiative. (2023). Green bonds market growth: A global overview. https://www.climatebonds.net

Climate Bonds Initiative. (2023). Green bonds market summary 2023. https://www.climatebonds.net/

Cobbinah, P. B., Amoateng, P., & Aboagye, E. (2015). Urbanization and the challenges of sustainable development in Africa. *American Journal of Rural Development*, *3*(6), 167–175. DOI: 10.12691/ajrd-3-6-1

Connelly, S. (2007). Mapping sustainable development as a contested concept. *Local Environment*, *12*(3), 259–278. DOI: 10.1080/13549830601183289

Convergence. (2023). Blended finance: How it works and why it matters. https://www.convergence.finance

Copenhagen Municipality. (2023). Copenhagen's climate action plan: Carbon neutrality by 2025. https://www.copenhagen.dk

Copenhagen Municipality. (2023). Copenhagen climate plan 2025. https://international.kk.dk/climate

Corporate Knights. (2023). Global 100 most sustainable corporations in the world. https://www.corporateknights.com/

Dalkmann, H., & Brannigan, C. (2007). Sustainable transport and urban mobility: The avoid-shift-improve approach. World Resources Institute. https://www.wri.org/avoid-shift-improve

Dávila, J. (2023). The urban cable car in Medellín: Reducing inequality through transport infrastructure. *Urban Mobility*, *18*(3), 105–120. DOI: 10.1016/j.umb.2023.03.007

Dávila, J. D. (2023). Urban mobility and social equity: Lessons from Medellín's Metrocable. *Urban Studies (Edinburgh, Scotland)*, *60*(2), 287–304. DOI: 10.1177/00420980221134556

Davis, J. E. (2010). *The community land trust reader*. Lincoln Institute of Land Policy.

Dempsey, N., Bramley, G., Power, S., & Brown, C. (2011). The social dimension of sustainable development: Defining urban social sustainability. *Sustainable Development*, *19*(5), 289–300. DOI: 10.1002/sd.417

Ellen MacArthur Foundation. (2023). Circular economy in urban environments: Innovations for sustainability. https://www.ellenmacarthurfoundation.org

Elmqvist, T., Andersson, E., McPhearson, T., & Norström, A. V. (2019). Sustainability and resilience for transformation in the urban century. *Nature Sustainability*, *2*(4), 267–273. DOI: 10.1038/s41893-019-0224-9

European Cycling Federation. (2023). The role of cycling in sustainable urban mobility: 2023 report. https://ecf.com/

European Cycling Federation. (2023). Cycling infrastructure in Copenhagen: A case study. https://www.ecf.com

European Cycling Federation. (2023). Cycling facts and figures: Copenhagen. https://ecf.com/

Fainstein, S. S. (2010). *The just city*. Cornell University Press.

Fernandez, G., Perez, J., & Vicente, P. (2019). Community-based disaster risk reduction: A case study from the Philippines. *Disaster Prevention and Management*, *28*(3), 306–322. DOI: 10.1108/DPM-08-2018-0247

Folke, C. (2006). Resilience: The emergence of a perspective for social–ecological systems analyses. *Global Environmental Change*, *16*(3), 253–267. DOI: 10.1016/j.gloenvcha.2006.04.002

Glaeser, E. L. (2011). *Triumph of the city: How our greatest invention makes us richer, smarter, greener, healthier, and happier*. Penguin Press.

Global Covenant of Mayors for Climate & Energy. (2023). Cities leading the way to a sustainable future. https://globalcovenantofmayors.org/

Greater London Authority. (2023). Ultra Low Emission Zone (ULEZ) impact report. https://www.london.gov.uk/what-we-do/research-and-analysis/environment/ultra-low-emission-zone

Hopkins, R. (2019). *From what is to what if: Unleashing the power of imagination to create the future we want*. Chelsea Green Publishing.

Intergovernmental Panel on Climate Change (IPCC). (2022). Climate change 2022: Impacts, adaptation, and vulnerability. https://www.ipcc.ch/report/ar6/wg2/

IPCC. (2022). Climate Change 2022: Impacts, Adaptation and Vulnerability. Intergovernmental Panel on Climate Change. https://www.ipcc.ch/report/ar6/wg2/

Kennedy, C., Cuddihy, J., & Engel-Yan, J. (2011). The study of urban metabolism and its applications to urban planning and design. *Environmental Pollution*, *159*(8–9), 1965–1973. DOI: 10.1016/j.envpol.2011.02.023

Kim, Y., & Kaneda, M. (2018). Post-earthquake building resilience: Lessons from Japan's building code reform after the 1995 Kobe earthquake. *International Journal of Disaster Risk Reduction*, *31*, 531–540. DOI: 10.1016/j.ijdrr.2018.07.004

Landrigan, P. J., Fuller, R., & Acosta, N. J. R.. (2018). The Lancet Commission on pollution and health. *Lancet*, *391*(10119), 462–512. DOI: 10.1016/S0140-6736(17)32345-0

Lefebvre, H. (1968). Le droit à la ville. *Anthropos: Studie z Oboru Anthropologie, Paleoethnologie, Paleontologie a Kvartérní Geologie*.

Li, X., Li, J., & Wang, E. (2020). Sponge city construction in China: A policy perspective. *Environmental Science & Policy*, *109*, 14–22. DOI: 10.1016/j.envsci.2020.03.008

Marchese, A. J., McPhearson, T., & Pickett, S. T. A. (2018). Resilience and sustainability in urban systems. *Urban Ecosystems, 21*(3), 423–438. DOI: 10.1007/s11252-018-0747-4

Medellín Municipality. (2023). Social urbanism and the transformation of Medellín. https://www.medellin.gov.co/

Meerow, S., Newell, J. P., & Stults, M. (2016). Defining urban resilience: A review. *Landscape and Urban Planning, 147*, 38–49. DOI: 10.1016/j.landurbplan.2015.11.011

Mueller, N., Rojas-Rueda, D., & Tainio, M. (2015). Health impact assessment of active transport: Benefits of cycling and walking in urban environments. *Environmental Health Perspectives, 123*(7), 640–647. DOI: 10.1289/ehp.1409091

Newman, P., & Kenworthy, J. (2021). *The end of automobile dependence: How cities are moving beyond car-based planning*. Island Press.

OECD. (2023c). *Governing cities: Bringing together urban and regional development*. OECD Publishing., DOI: 10.1787/9789264291943-

Olsson, P., Galaz, V., & Boonstra, W. J. (2015). Sustainability transformations: A resilience perspective. *Ecology and Society, 20*(3), 1–19. DOI: 10.5751/ES-07887-200341

Organisation for Economic Co-operation and Development. (2023a). Cities policy responses to COVID-19: OECD report. OECD. https://www.oecd.org/coronavirus/policy-responses/cities-policy-responses-to-covid-19-df3f6e1d/

Organisation for Economic Co-operation and Development (OECD). (2023). Affordable housing and urban inequality: OECD statistics. https://www.oecd.org/urban/affordable-housing

Organisation for Economic Co-operation and Development (OECD). (2023). Urban governance and the challenges of fragmentation: OECD report. https://www.oecd.org

Pauleit, S., Zölch, T., Hansen, R., Randrup, T. B., & Konijnendijk van den Bosch, C. (2017). Nature-based solutions and climate change–four shades of green. Nature-based solutions to climate change adaptation in urban areas: Linkages between science, policy and practice, 29-49.

Ramaswami, A., Hillman, T., Janson, B., Reiner, M., & Thomas, G. (2017). A social–ecological approach to the urban metabolism of cities. *Nature Sustainability*, *1*(5), 213–222. DOI: 10.1038/s41893-018-0033-3

Raworth, K. (2017). *Doughnut economics: Seven ways to think like a 21st-century economist*. Chelsea Green Publishing.

Resilient Cities Network. (2023). Urban resilience for whom, what, when, where, and why? https://www.resilientcitiesnetwork.org/urban-resilience

Rolnik, R., Giraudo, A., & Cardoso, A. (2021). Minha Casa Minha Vida: The challenges of large-scale affordable housing in Brazil. *Housing Studies*, *36*(1), 123–145. DOI: 10.1080/02673037.2020.1789642

Rutherford, J. (2014). The Vicissitudes of District Energy: Consolidation, loss and disconnection in Stockholm's district heating grid. *Urban Studies (Edinburgh, Scotland)*, *51*(5), 1055–1071. DOI: 10.1177/0042098013495573

Rutherford, L. (2014). Decentralized governance and sustainable urban development in Sweden and Switzerland. *Sustainable Cities and Society*, *12*, 38–49. DOI: 10.1016/j.scs.2014.03.001

Sachs, J. D., Schmidt-Traub, G., Kroll, C., Lafortune, G., & Fuller, G. (2021). *Sustainable development report 2021: The Decade of Action for the Sustainable Development Goals*. Cambridge University Press.

Sassen, S. (2018). *Expulsions: Brutality and complexity in the global economy*. Belknap Press.

Singapore, P. U. B. (2023). Water management and sustainability in Singapore: The Four National Taps policy. https://www.pub.gov.sg

Singapore, P. U. B. (2023). Singapore's water story: Four national taps. https://www.pub.gov.sg/

Singapore Government Technology Agency. (2023). AI-based waste management in Singapore: Optimizing urban sustainability. https://www.tech.gov.sg

Singapore Urban Redevelopment Authority. (2023). Smart and sustainable urban design in Singapore. https://www.ura.gov.sg

Singapore Urban Redevelopment Authority. (2023). Urban sustainability and smart nation initiatives. https://www.ura.gov.sg/

Smart Nation Singapore. (2023). The Smart Nation initiative: Using technology for sustainable urban development. https://www.smartnation.gov.sg

Smart Nation Singapore. (2023). What is Smart Nation? https://www.smartnation.gov.sg/

Soja, E. W. (2010). *Seeking spatial justice*. University of Minnesota Press.

State of Green. (2023). Copenhagen's smart energy system: A sustainable model for the future. https://stateofgreen.com/en/partners/copenhagen

State of Green. (2023). Copenhagen's renewable energy initiatives and district heating system. https://stateofgreen.com/en/partners/copenhagen

State of Green. (2023). Denmark's green transition: Leading by example. https://stateofgreen.com/

Stickler, L. (2022). The economic costs of urban traffic congestion. *Urban Economics Journal*, *12*(4), 456–470.

Tesla. (2023). Energy products: Powerpack and Powerwall. https://www.tesla.com/energy

Tummers, L. (2016). Innovative housing strategies for sustainable cities: Adaptive reuse, incremental housing, and co-housing. *Sustainable Development*, *24*(6), 326–340. DOI: 10.1002/sd.1602

UN-Habitat. (2020). World Cities Report 2020: The value of sustainable urbanization. United Nations Human Settlements Programme. https://unhabitat.org/wcr/

UN-Habitat. (2023). *World cities report 2022: Envisaging the future of cities*. United Nations Human Settlements Programme.

UN-Habitat. (2023). World cities report 2023: The value of sustainable urbanization. https://unhabitat.org/world-cities-report-2023

UN-Habitat. (2023). World cities report 2023: The value of sustainable urbanization. https://unhabitat.org/world-cities-report-2023

UNESCO. (2022). Culture and sustainable development. https://en.unesco.org/themes/culture-and-sustainable-development

United Nations Department of Economic and Social Affairs. (2023). *World population prospects 2022: Summary of results*. United Nations.

United Nations Department of Economic and Social Affairs. (2023). *Goal 11: Sustainable cities and communities*. United Nations.

United Nations Department of Economic and Social Affairs. (2023). World population prospects 2022: Summary of results. United Nations. https://population.un.org/wpp/Publications/Files/WPP2022_Summary-of-Results.pdf

United Nations Department of Economic and Social Affairs. (2023). World urbanization prospects 2022: Highlights. United Nations. https://population.un.org/wup/Publications/Files/WUP2022_Highlights.pdf

United Nations Human Settlements Programme (UN-Habitat). (2020). Urban governance for sustainable development. https://unhabitat.org

United Nations Office for Disaster Risk Reduction (UNDRR). (2015). Sendai Framework for Disaster Risk Reduction 2015–2030. https://www.undrr.org/implementing-sendai-framework

Wirth, P., Garschagen, M., & Cloke, H. (2022). Urban resilience transformation in Medellín: Insights from twenty years of social urbanism. *Sustainability*, *14*(2), 891. DOI: 10.3390/su14020891

Wolman, A. (1965). The metabolism of cities. *Scientific American*, *213*(3), 179–190. DOI: 10.1038/scientificamerican0965-179

World Bank. (2022). The world bank annual report 2022: Financing development through a green, resilient, and inclusive recovery. World Bank Group. https://www.worldbank.org/en/publication/annual-report

World Bank. (2023). Financing sustainable cities: Challenges and opportunities for SDG 11. https://www.worldbank.org/en/topic/urbandevelopment

World Business Council for Sustainable Development. (2023). Business solutions for sustainable cities. https://www.wbcsd.org/

World Health Organization. (2023). Air pollution levels rising in many of the world's poorest cities. https://www.who.int/news/item/29-09-2023-air-pollution-levels-rising-in-many-of-the-world's-poorest-cities

World Health Organization (WHO). (2023). Air pollution and health. https://www.who.int/health-topics/air-pollution

Compilation of References

Abbas, A., Luo, X., Wattoo, M. U., & Hu, R. X. (2022). Organizational Behavior in Green Supply Chain Integration: Nexus Between Information Technology Capability, Green Innovation, and Organizational Performance. *Frontiers in Psychology*, *13*, 874639. Advance online publication. DOI: 10.3389/fpsyg.2022.874639 PMID: 35391968

Abbas, J., Balsalobre-Lorente, D., Amjid, M. A., Al-Sulaiti, K., Al-Sulaiti, I., & Aldereai, O. (2024). Financial innovation and digitalization promote business growth: The interplay of green technology innovation, product market competition and firm performance. *Innovation and Green Development*, *3*(1), 100111. DOI: 10.1016/j.igd.2023.100111

Abrego, D., Sánchez-Gutiérrez, J., & Medina-Quintero, J. M. (2018). Influence of information systems on organizational results. *Contaduría y Administración*, *63*(1), 1–20. DOI: 10.22201/fca.24488410e.2018.1256

Acevedo, B., & Malevicius, R. (2021). Rethinking Education for Sustainability in Management Education: Going beyond Metrics toward Human Virtues. In *Handbook of Research on International Business and Models for Global Purpose-Driven Companies* (pp. 219–234). IGI Global. DOI: 10.4018/978-1-7998-4909-4.ch012

Acuña, N., Figueroa, L., & Wilches, M. J. (2017). Influencia de los Sistemas de Gestión Ambiental ISO 14001 en las organizaciones: Caso estudio empresas manufactureras de Barranquilla. *Ingeniare. Revista Chilena de Ingeniería*, *25*(1), 143–153. DOI: 10.4067/S0718-33052017000100143

Adadi, A., & Berrada, M. (2018). Peeking inside the black-box: A survey on explainable artificial intelligence (XAI). *IEEE Access : Practical Innovations, Open Solutions*, *6*, 52138–52160. DOI: 10.1109/ACCESS.2018.2870052

Agyeman, J., Bullard, R. D., & Evans, B. (2016). *Just sustainabilities: Development in an unequal world*. MIT Press.

AI in accounting: benefits and associated risks. (2023, November 20). Retrieved from https://incp.org.co/publicaciones/boletin-virtual/2023/11/la-ia-en-la-contabilidad-beneficios-y-riesgos-asociados/

Alhassan, I., Sammon, D., & Daly, M. (2019). Critical success factors for data governance: A telecommunications case study. *Journal of Decision Systems*, *28*(1), 41–61. DOI: 10.1080/12460125.2019.1633226

Alhosani, K., & Alhashmi, S. M. (2024). Opportunities, challenges, and benefits of AI innovation in government services: A review. *Discover Artificial Intelligence*, *4*(1), 18. DOI: 10.1007/s44163-024-00111-w

Amazon and Big Data. A success story. (n.d.). Retrieved from https://ayudaleyprotecciondatos.es/big-data/amazon/

Angarita, G. T. (2023). *The impacts of artificial intelligence on the role of the public accountant*. Faculty of Economic, Administrative and Accounting Sciences, Cooperative University.

Anthony, B.Jr. (2019). Green information system integration for environmental performance in organizations: An extension of belief-action-outcome framework and natural resource-based view theory. *Benchmarking*, *26*(3), 1033–1062. DOI: 10.1108/BIJ-05-2018-0142

Arslan, S. (2024, May 28). Bitcoin Price Prediction Using Sentiment Analysis and Empirical Mode Decomposition. *Computational Economics*, 1–22.

Asheim, G. B. Sustainability: Ethical Foundations and Economic Properties; Policy Research Working Paper Series 1302; The World Bank: Washington, DC, USA, 1994.

Attaran, M. (2020). Digital technology enablers and their implications for supply chain management. *Supply Chain Forum: An International Journal (21)*3, 158-172. DOI: 10.1080/16258312.2020.1751568

Bachmann, N., Tripathi, S., Brunner, M., & Jodlbauer, H. (2022). The contribution of data-driven technologies in achieving the sustainable development goals. *Sustainability (Basel)*, *14*(5), 2497. DOI: 10.3390/su14052497

Baggia, A., Maletič, M., Žnidaršič, A., & Brezavšček, A. (2019). Drivers and outcomes of green IS adoption in small and medium-sized enterprises. *Sustainability (Basel)*, *11*(6), 1575. DOI: 10.3390/su11061575

Bai, X., Roberts, B., & Chen, J. (2016). Urban sustainability experiments in Asia: Patterns and pathways. *Environmental Science & Policy*, *62*, 1–12. DOI: 10.1016/j.envsci.2016.01.008

Bansal, V., & Agarwal, A. (2015). Enterprise resource planning: Identifying relationships among critical success factors. *Business Process Management Journal*, *21*(6), 1337–1352. DOI: 10.1108/BPMJ-12-2014-0128

Barcelona Digital City. (2023). Smart city traffic management in Barcelona: A case study. https://www.barcelona.cat/digitalcity

Barocas, S., Hardt, M., & Narayanan, A. (2019). *Fairness and machine learning*. http://fairmlbook.org

Barroso, L. A., & Hölzle, U. (2007). The Case for Energy-Proportional Computing. *Computer*, *40*(12), 33–37. DOI: 10.1109/MC.2007.443

Beatley, T. (2012). *Green urbanism: Learning from European cities*. Island Press.

Bender, E. M., Gebru, T., McMillan-Major, A., & Shmitchell, S. On the Dangers of Stochastic Parrots: Can Language Models Be Too Big? In *Proceedings of the 2021 ACM Conference on Fairness, Accountability, and Transparency, Virtual Event*, 3–10 March 2021; pp. 610–623. DOI: 10.1145/3442188.3445922

Bengio, Y., Delalleau, O., & Le Roux, N. (2005). The Curse of Dimensionality for Local Kernel Machines. *Techn. Rep., 1258*, 1.

Benhmad, T., Rhaimi, C. B., Alomari, S., & Aljuhani, L. (2024). Design and Implementation of an Integrated IoT and Artificial Intelligence System for Smart Irrigation Management. *International Journal of Advances in Soft Computing & Its Applications*, *16*(1).

Berti-Equille, L., Dao, D., Ermon, S., & Goswami, B. Challenges in KDD and ML for Sustainable Development. In *Proceedings of the 27th ACM SIGKDD Conference on Knowledge Discovery & Data Mining*, Singapore, 14 August 2021; pp. 4031–4032. DOI: 10.1145/3447548.3470798

Bexell, M., & Jönsson, K. (2017). Responsibility and the United Nations' sustainable development goals. *Forum for Development Studies*, *44*(1), 13–29. DOI: 10.1080/08039410.2016.1252424

Bhati, R., & Mittal, S. (2023). The Role and Impact of Artificial Intelligence in Attaining Sustainability Goals. In *2023 9th International Conference on Advanced Computing and Communication Systems (ICACCS)*. DOI: 10.1109/ICACCS57279.2023.10112887

Biermann, F., Kanie, N., & Kim, R. E. (2017). Global governance by goal-setting: The novel approach of the UN sustainable development goals. *Current Opinion in Environmental Sustainability*, *26–27*, 26–31. DOI: 10.1016/j.cosust.2017.01.010

Biggi, G., Mina, A., & Tamagni, F. (2023). There are different shades of green: heterogeneous environmental innovations and their effects on firm performance. *arXiv preprint arXiv:2310.08353*.Disponible en: https://arxiv.org/abs/2310.08353

Binns, R. (2018). Fairness in machine learning: Lessons from political philosophy. *Proceedings of the 2018 Conference on Fairness, Accountability, and Transparency*, 149-159. DOI: 10.1145/3287560.3287576

Bin, Y., Yang, Y., Shen, F., Xie, N., Shen, H. T., & Li, X. (2019, July). Describing video with attention-based bidirectional LSTM. *IEEE Transactions on Cybernetics*, *49*(7), 2631–2641. DOI: 10.1109/TCYB.2018.2831447 PMID: 29993730

Boden, M. A. (2018). *Artificial intelligence: A very short introduction*. Oxford University Press. DOI: 10.1093/actrade/9780199602919.001.0001

Bolte, L., Vandemeulebroucke, T., & Van Wynsberghe, A. (2022). From an Ethics of Carefulness to an Ethics of Desirability: Going Beyond Current Ethics Approaches to Sustainable AI. *Sustainability (Basel)*, *14*(8), 4472. DOI: 10.3390/su14084472

Braun, V., & Clarke, V. (2006). Using thematic analysis in psychology. *Qualitative Research in Psychology*, *3*(2), 77–101. DOI: 10.1191/1478088706qp063oa

Brevini, B. (2021). *Is AI Good for the Planet?* Polity Press. [Google Scholar]

Brown, R. E., Brown, R., Masanet, E., Nordman, B., Tschudi, B., Shehabi, A., Stanley, J., Koomey, J., Sartor, D., & Chan, P. (2007). *Report to Congress on Server and Data Center Energy Efficiency: Public Law 109-431; Lawrence Berkeley National Laboratory*. LBNL.

Brundtland, G. (1987). Our Common Future—Call for Action. *Environmental Conservation*, *14*(4), 291–294. DOI: 10.1017/S0376892900016805

Bukhari, A. H., Raja, M. A. Z., Sulaiman, M., Islam, S., Shoaib, M., & Kumam, P. (2020, April). Fractional neuro-sequential ARFIMA-LSTM for financial market forecasting. *IEEE Access : Practical Innovations, Open Solutions*, *8*, 71326–71338. DOI: 10.1109/ACCESS.2020.2985763

Bullard, R. D. (1993). *Confronting Environmental Racism: Voices from the Grassroots*. South End Press.

Bullen, C. V., & Rockart, J. F. (1981). A primer on critical success factors. http://mit.dspace.org/bitstream/handle/1721.1/1988/SWP-1220-08368993 - CISR-069.pdf?sequence=1

Buolamwini, J., & Gebru, T. (2018). Gender shades: Intersectional accuracy disparities in commercial gender classification. *Proceedings of Machine Learning Research, 81*, 1–15.

Burström, T., Parida, V., Lahti, T., & Wincent, J. (2021). AI-enabled business-model innovation and transformation in industrial ecosystems: A framework, model and outline for further research. *Journal of Business Research, 127*, 85–95. DOI: 10.1016/j.jbusres.2021.01.016

Bustani, P., Mulyanto, S., & Nurjanana (2018). Prediction of electricity sales using neural based inverse distance weighting method. *International Journal of Engineering & Technology, 7*(2.2), 65-69.

Bustos, E. V. (2017). *Digital marketing*.

Butler, T. (2011). Compliance with institutional imperatives on environmental sustainability: Building theory on the role of Green IS. Journal of strategic information systems, 20(1, SI), 6–26. https://doi.org/DOI: 10.1016/j.jsis.2010.09.006

C40 Cities. (2023). Cities and climate change: A global perspective. https://www.c40.org

C40 Cities. (2023). Cities are responsible for 70% of global greenhouse gas emissions. https://www.c40.org/researches/cities-are-responsible-for-70-of-global-greenhouse-gas-emissions

C40 Cities. (2023). How cities are leading the way in climate action. https://www.c40.org/

C40 Cities. (2023). The impact of Bus Rapid Transit systems in major global cities. https://www.c40.org/initiatives/bus-rapid-transit

Cabannes, Y. (2015). Participatory budgeting: A powerful, yet underutilized, local governance tool for municipalities. UN-Habitat. https://unhabitat.org/participatory-budgeting-a-powerful-yet-underutilized-local-governance-tool-for-municipalities

Cai, Y. J., & Lo, C. K. (2020). Omni-channel management in the new retailing era: A systematic review and future research agenda. *International Journal of Production Economics, 229*, 107729. DOI: 10.1016/j.ijpe.2020.107729

Calavita, N., & Mallach, A. (2010). *Inclusionary housing in international perspective: Affordable housing, social inclusion, and land use regulation*. Policy Press.

Campbell, S. (1996). Green cities, growing cities, just cities?: Urban planning and the contradictions of sustainable development. *Journal of the American Planning Association, 62*(3), 296–312. DOI: 10.1080/01944369608975696

Cantú-Ortiz, F. J., Galeano Sánchez, N., Garrido, L., Terashima-Marin, H., & Brena, R. F. (2020). An artificial intelligence educational strategy for the digital transformation. [IJIDeM]. *International Journal on Interactive Design and Manufacturing, 14*(4), 1195–1209. DOI: 10.1007/s12008-020-00702-8

Cao, T., Lim, E.-P., Zhou, Z.-H., Ho, T.-B., Cheung, D., & Motoda, H. (Eds.). Springer International Publishing: Cham, Switzerland, 2015; Volume 9078, pp. 27–38. ISBN 978-3-319-18031-1.

Cardoso, A., & Jaenisch, P. (2017). Challenges in Brazil's affordable housing program: Urbanity, location, and social segregation. *Urban Studies (Edinburgh, Scotland), 54*(10), 2289–2308. DOI: 10.1177/0042098016680919

Castán Broto, V., & Westman, L. (2020). Just transitions and the co-production of urban climate solutions. *Urban Studies (Edinburgh, Scotland), 57*(7), 1307–1323. DOI: 10.1177/0042098019880670

Chen, B., Qiao, S., Zhao, J., Liu, D., Shi, X., Lyu, M., Chen, H., Lu, H., & Zhai, Y. (2021). A security awareness and protection system for 5G smart healthcare based on zero-trust architecture. *IEEE Internet of Things Journal, 8*(13), 10466–10476. DOI: 10.1109/JIOT.2020.3041042 PMID: 35783535

Chen, C. S., Yu, C. C., & Hu, J. S. (2018). Constructing performance measurement indicators to suggested corporate environmental responsibility framework. *Technological Forecasting and Social Change*, *135*, 33–43. DOI: 10.1016/j.techfore.2017.05.033

Chen, C.-J., Huang, Y.-Y., Li, Y.-S., Chang, C.-Y., & Huang, Y.-M. (2020). An AIoT Based Smart Agricultural System for Pests Detection. *IEEE Access : Practical Innovations, Open Solutions*, *8*, 180750–180761. DOI: 10.1109/ACCESS.2020.3024891

Chen, Y., Zhang, L., & Li, X. (2014). Compressed urbanization and its impact on sustainable development in China. *Sustainable Cities and Society*, *13*, 1–8. DOI: 10.1016/j.scs.2014.03.003

Chui, M., Manyika, J., & Miremadi, M. (2021). *The impact of AI on work: An economy-wide view*. McKinsey Global Institute.

Chunque, V. (2021). *Digital marketing and positioning level of the company*.

Ciarli, T., Kenney, M., Massini, S., & Piscitello, L. (2021). Digital technologies, innovation, and skills: Emerging trajectories and challenges. *Research Policy*, *50*(7), 104289. DOI: 10.1016/j.respol.2021.104289

City of Amsterdam. (2023). *The Amsterdam City Doughnut: A tool for transformative action*. Circle Economy.

Clark, J. A., Liu, Y., & Isaias, P. (2020). Critical success factors for implementing learning analytics in higher education: A mixed-method inquiry. *Australasian Journal of Educational Technology*, *36*(6), 89–106. DOI: 10.14742/ajet.6164

Climate Bonds Initiative. (2023). Green bonds market growth: A global overview. https://www.climatebonds.net

Climate Bonds Initiative. (2023). Green bonds market summary 2023. https://www.climatebonds.net/

Cobbinah, P. B., Amoateng, P., & Aboagye, E. (2015). Urbanization and the challenges of sustainable development in Africa. *American Journal of Rural Development*, *3*(6), 167–175. DOI: 10.12691/ajrd-3-6-1

Coeckelbergh, M. (2021). AI for Climate: Freedom, Justice, and Other Ethical and Political Challenges. *AI and Ethics*, *1*(1), 67–72. DOI: 10.1007/s43681-020-00007-2

Connelly, S. (2007). Mapping sustainable development as a contested concept. *Local Environment*, *12*(3), 259–278. DOI: 10.1080/13549830601183289

Convergence. (2023). Blended finance: How it works and why it matters. https://www.convergence.finance

Cooper, R. G. (2024). The AI transformation of product innovation. *Industrial Marketing Management*, *119*, 62–74. DOI: 10.1016/j.indmarman.2024.03.008

Copenhagen Municipality. (2023). Copenhagen climate plan 2025. https://international.kk.dk/climate

Copenhagen Municipality. (2023). Copenhagen's climate action plan: Carbon neutrality by 2025. https://www.copenhagen.dk

Corporate Knights. (2023). Global 100 most sustainable corporations in the world. https://www.corporateknights.com/

Coulson, R. N., Folse, L. J., & Loh, D. K. (1987). Artificial intelligence and natural resource management. *Science*, *237*(4812), 262–267. DOI: 10.1126/science.237.4812.262 PMID: 17772054

Cowls, J., Tsamados, A., Taddeo, M., & Floridi, L. (2023). The AI Gambit: Leveraging Artificial Intelligence to Combat Climate Change—Opportunities, Challenges, and Recommendations. *AI & Society*, *38*(1), 283–307. DOI: 10.1007/s00146-021-01294-x PMID: 34690449

Crawford, K., & Joler, V. (2018). Anatomy of an AI system: The Amazon Echo as an anatomical map of human labor, data, and planetary resources. *AI Now Institute*. https://anatomyof.ai

Crawford, K. (2021). *The Atlas of AI: Power, Politics, and the Planetary Costs of Artificial Intelligence*. Yale University Press.

Creswell, J. W. (2014). *Research design: Qualitative, quantitative, and mixed methods approaches* (4th ed.). SAGE Publications.

Cui, S., Gao, Y., Huang, Y., Shen, L., Zhao, Q., Pan, Y., & Zhuang, S. (2023). Advances and Applications of Machine Learning and Deep Learning in Environmental Ecology and Health. *Environmental Pollution, 335*, 122358. DOI: 10.1016/j.envpol.2023.122358 PMID: 37567408

D'Silva, D., & Ambawade, D. D. (2021). Building a zero trust architecture using Kubernetes. In *Proceedings of the 6th International Conference for Convergence in Technology (I2CT)*. DOI: 10.1109/I2CT51068.2021.9418203

Dalkmann, H., & Brannigan, C. (2007). Sustainable transport and urban mobility: The avoid-shift-improve approach. World Resources Institute. https://www.wri.org/avoid-shift-improve

Daniel, D. R. (1961). Management information crisis. *Harvard Business Review*, •••, 111–121.

Danso, A., Adomako, S., Amankwah-Amoah, J., Owusu-Agyei, S., & Konadu, R. (2019). Environmental sustainability orientation, competitive strategy, and financial performance. *Business Strategy and the Environment, 28*(5), 885–895. DOI: 10.1002/bse.2291

Dávila, J. (2023). The urban cable car in Medellín: Reducing inequality through transport infrastructure. *Urban Mobility, 18*(3), 105–120. DOI: 10.1016/j.umb.2023.03.007

Dávila, J. D. (2023). Urban mobility and social equity: Lessons from Medellín's Metrocable. *Urban Studies (Edinburgh, Scotland), 60*(2), 287–304. DOI: 10.1177/00420980221134556

Davis, J. E. (2010). *The community land trust reader*. Lincoln Institute of Land Policy.

de Sousa, W. G., Khosravi, S., Cristoni, R., & Ketelhodt, L. (2021). The role of AI-powered drones in sustainable development: Evidence from reforestation projects. *Sustainability, 13*(15), 8505. DOI: 10.3390/su13158505

Dempsey, N., Bramley, G., Power, S., & Brown, C. (2011). The social dimension of sustainable development: Defining urban social sustainability. *Sustainable Development*, *19*(5), 289–300. DOI: 10.1002/sd.417

Denolf, J. M., Trienekens, J. H., Wognum, P. M., Van Der Vorst, J. G. A. J., & Omta, S. W. F. (2015). Towards a framework of critical success factors for implementing supply chain information systems. *Computers in Industry*, *68*, 16–26. DOI: 10.1016/j.compind.2014.12.012

Dhar, P. (2020). The carbon impact of artificial intelligence. *Nature Machine Intelligence*, *2*(8), 423–425. DOI: 10.1038/s42256-020-0219-9

Dilkina, B., Houtman, R., Gomes, C. P., Montgomery, C. A., McKelvey, K. S., Kendall, K., Graves, T. A., Bernstein, R., & Schwartz, M. K. (2017). Trade-offs and Efficiencies in Optimal Budget-constrained Multispecies Corridor Networks. [Google Scholar] [CrossRef]. *Conservation Biology*, *31*(1), 192–202. DOI: 10.1111/cobi.12814 PMID: 27677418

Dong, G., Li, R., Jiang, J., Wu, H., & McClure, S. C. (2020, September). Multigranular wavelet decomposition-based support vector regression and moving average method for service-time prediction on web map service platforms. *IEEE Systems Journal*, *14*(3), 3653–3664. DOI: 10.1109/JSYST.2019.2944527

Doroslovački, K., Gradojevic, N., & Tarnaud, A. C. (2024, June 11). A Novel Market Sentiment Analysis Model for Forecasting Stock and Cryptocurrency Returns. *IEEE Transactions on Systems, Man, and Cybernetics. Systems*, *54*(9), 5248–5259. DOI: 10.1109/TSMC.2024.3402160

Du Plooy, H., Tommasi, F., Furlan, A., Nenna, F., Gamberini, L., Ceschi, A., & Sartori, R. (2024). A human-centered perspective on individual risks for digital innovation management: An integrative conceptual review. *European Journal of Innovation Management*, *28*(11), 49–76. DOI: 10.1108/EJIM-09-2023-0821

Eikebrokk, T. R., & Iden, J. (2017). Strategising IT service management through ITIL implementation: Model and empirical test. *Total Quality Management & Business Excellence*, *28*(3-4), 238–265. DOI: 10.1080/14783363.2015.1075872

Eli-Chukwu, N. C. (2019). Applications of artificial intelligence in agriculture: A review. *Engineering, Technology & Applied Science Research*, *9*(4), 4. DOI: 10.48084/etasr.2756

Ellen MacArthur Foundation. (2023). Circular economy in urban environments: Innovations for sustainability. https://www.ellenmacarthur foundation.org

Elmqvist, T., Andersson, E., McPhearson, T., & Norström, A. V. (2019). Sustainability and resilience for transformation in the urban century. *Nature Sustainability*, *2*(4), 267–273. DOI: 10.1038/s41893-019-0224-9

Ennouri, K., Smaoui, S., Gharbi, Y., Cheffi, M., Ben Braiek, O., Ennouri, M., & Triki, M. A. (2021). Usage of artificial intelligence and remote sensing as efficient devices to increase agricultural system yields. *Journal of Food Quality*, *2021*(1), 6242288. DOI: 10.1155/2021/6242288

European Cycling Federation. (2023). Cycling facts and figures: Copenhagen. https://ecf.com/

European Cycling Federation. (2023). Cycling infrastructure in Copenhagen: A case study. https://www.ecf.com

European Cycling Federation. (2023). The role of cycling in sustainable urban mobility: 2023 report. https://ecf.com/

European Parliament. (2016). *Regulation (EU) 2016/679 of the European Parliament and of the Council of 27 April 2016*. General Data Protection Regulation.

Evans, R., & Gao, J. (2016). *DeepMind AI reduces Google data centre cooling bill by 40*. DeepMind.

Fainstein, S. S. (2010). *The just city*. Cornell University Press.

Fan, C., Wang, J., Gang, W., & Li, S. (2019). Assessment of deep recurrent neural network -based strategies for short-term building energy predictions. *Applied Energy*, *236*, 700–710. DOI: 10.1016/j.apenergy.2018.12.004

Faul, F., Erdfelder, E., Buchner, A., & Lang, A. G. (2009). Statistical power analyses using G* Power 3.1: Tests for correlation and regression analyses. *Behavior Research Methods, 41*(4), 1149–1160. DOI: 10.3758/BRM.41.4.1149 PMID: 19897823

Felicetti, A. M., Corvello, V., & Ammirato, S. (2024). Digital innovation in entrepreneurial firms: A systematic literature review. *Review of Managerial Science, 18*(2), 315–362. DOI: 10.1007/s11846-023-00638-9

Feng, M., Yu, W., Wang, X., Wong, C. Y., Xu, M., & Xiao, Z. (2018). Green supply chain management and financial performance: The mediating roles of operational and environmental performance. *Business Strategy and the Environment, 27*(7), 811–824. DOI: 10.1002/bse.2033

Feng, M., Zheng, J., Ren, J., Hussain, A., Li, X., Xi, Y., & Liu, Q. (2019, July). Big data analytics and mining for effective visualization and trends forecasting of crime data. *IEEE Access : Practical Innovations, Open Solutions, 7*, 106111–106123. DOI: 10.1109/ACCESS.2019.2930410

Fernandez, G., Perez, J., & Vicente, P. (2019). Community-based disaster risk reduction: A case study from the Philippines. *Disaster Prevention and Management, 28*(3), 306–322. DOI: 10.1108/DPM-08-2018-0247

Finio, M., & D., A. (2023, December 8). *Artificial intelligence in finance*. IBM. Retrieved from https://www.ibm.com/mx-es/topics/artificial-intelligence-finance

Fisher, D. A Selected Summary of AI for Computational Sustainability. In *Proceedings of the Thirty-First AAAI Conference on Artificial Intelligence (AAAI-17)*, San Francisco, CA, USA, 4–9 February 2017. [Google Scholar] DOI: 10.1609/aaai.v31i1.11148

Floridi, L., Cowls, J., Beltrametti, M., Chatila, R., Chazerand, P., Dignum, V., Luetge, C., Madelin, R., Pagallo, U., Rossi, F., Schafer, B., Valcke, P., & Vayena, E. (2018). AI4People—An ethical framework for a good AI society: Opportunities, risks, principles, and recommendations. *Minds and Machines, 28*(4), 689–707. DOI: 10.1007/s11023-018-9482-5 PMID: 30930541

Folke, C. (2006). Resilience: The emergence of a perspective for social–ecological systems analyses. *Global Environmental Change*, *16*(3), 253–267. DOI: 10.1016/j.gloenvcha.2006.04.002

Fonseca, P. G., Souza, D. M. O. R., Santos, E. M., & Miranda, M. A. S. (2019). Fatores críticos de sucesso na implementação de sistemas de informação a partir da percepção de usuários: Uma experiência no setor público. *Holos*, *35*(3), 1–19. DOI: 10.15628/holos.2019.8190

Fornell, C., & Larcker, D. F. (1981). Structural equation models with unobservable variables and measurement error. Algebra and statistics. https://doi.org/DOI: 10.1177/002224378101800313

Franco Valdez, A. (2024). *The implementation of artificial intelligence in retail commerce*. Retrieved from https://centrocomerciodetallista.tec.mx/sites/g/files/vgjovo1571/files/CCD_Inteigencia%20Artificial%2003

Franganillo. (2023). *The large language models. ThinKEPI Yearbook, 17*.

Ganaie, M. A., Hu, M., Malik, A. K., Tanveer, M., & Suganthan, P. N. (2022). Ensemble Deep Learning: A Review. *Engineering Applications of Artificial Intelligence*, *115*, 105151. DOI: 10.1016/j.engappai.2022.105151

Garbin, C., Zhu, X., & Marques, O. (2020). Dropout vs. Batch Normalization: An Empirical Study of Their Impact to Deep Learning. *Multimedia Tools and Applications*, *79*(19-20), 12777–12815. DOI: 10.1007/s11042-019-08453-9

García, J. D., & Rodríguez, M. A. (2022). Factores críticos de éxito en la implementación de un ERP en una empresa del sector de alimentos. *Trabajo de grado, Universidad Nacional de Colombia.*https://repositorio.unal.edu.co/bitstream/handle/unal/81260/74130731.2022.pdf

Garg, P., & Agarwal, D. (2014). Critical success factors for ERP implementation in a Fortis hospital: An empirical investigation. *Journal of Enterprise Information Management*, *27*(4), 402–423. DOI: 10.1108/JEIM-06-2012-0027

Garmendia, L. S. (2024, March 15). *Beyond the numbers: This is the AI that will take the financial sector into the future*. Retrieved from https://forbes.com.mx/mas-alla-de-los-numeros-asi-es-la-ia-que-llevara-al-sector-financiero-al-futuro/

Gebru, T., Morgenstern, J., Vecchione, B., Vaughan, J. W., Wallach, H., Daumé, H.III, & Crawford, K. (2021). Datasheets for datasets. *Communications of the ACM*, *64*(12), 86–92. DOI: 10.1145/3458723

Gholami, R., Molla, A., Goswami, S., & Brewster, C. (2018). Green information systems use in social enterprise: The case of a community-led eco-localization website in the West Midlands region of the UK. *Information Systems Frontiers*, *20*(6), 1345–1361. https://link.springer.com/article/10.1007/s10796-016-9733-z. DOI: 10.1007/s10796-016-9733-z

Glaeser, E. L. (2011). *Triumph of the city: How our greatest invention makes us richer, smarter, greener, healthier, and happier*. Penguin Press.

Glaeser, S., & Lang, M. (2024). Measuring innovation and navigating its unique information issues: A review of the accounting literature on innovation. *Journal of Accounting and Economics*, *78*(2-3), 101720. DOI: 10.1016/j.jacceco.2024.101720

Global Covenant of Mayors for Climate & Energy. (2023). Cities leading the way to a sustainable future. https://globalcovenantofmayors.org/

Goodman, A. (2019, June 25). *What is Zero Trust?* DZone. Retrieved from https://dzone.com/articles/what-is-zero-trust

Gorla, N., Somers, T. M., & Wong, B. (2010). Organizational impact of system quality, information quality and service quality. *The Journal of Strategic Information Systems*, *19*(3), 207–228. DOI: 10.1016/j.jsis.2010.05.001

Greater London Authority. (2023). Ultra Low Emission Zone (ULEZ) impact report. https://www.london.gov.uk/what-we-do/research-and-analysis/environment/ultra-low-emission-zone

Green, K.Jr, Zelbst, P. J., Meacham, J., & Bhadauria, V. S. (2012a). Green supply chain management practices: Impact on performance. *Supply Chain Management, 17*(3), 290–305. DOI: 10.1108/13598541211227126

Green, K.Jr, Zelbst, P., Bhadauria, V., & Meacham, J. (2012b). Do environmental collaboration and monitoring enhance organizational performance? *Industrial Management & Data Systems, 112*(2), 186–205. DOI: 10.1108/02635571211204254

Guo, X., Li, M., Wang, Y., & Mardani, A. (2023). Does digital transformation improve the firm's performance? From the perspective of digitalization paradox and managerial myopia. *Journal of Business Research, 163*, 113868. DOI: 10.1016/j.jbusres.2023.113868

Haefner, N., Wincent, J., Parida, V., & Gassmann, O. (2021). Artificial intelligence and innovation management: A review, framework, and research agenda . *Technological Forecasting and Social Change, 162*, 120392. DOI: 10.1016/j.techfore.2020.120392

Hair, J. F., Hult, G. T. M., Ringle, C. M., & Sarstedt, M. (2017). *A Primer on Partial Least Squares Structural Equation Modeling (PLS-SEM)* (2nd ed.). Sage.

Hanin, B. Which Neural Net Architectures Give Rise to Exploding and Vanishing Gradients? In *Proceedings of the 32nd Conference on Neural Information Processing Systems (NeurIPS 2018)*, Montréal, Canada, 2–8 December 2018; Volume 31.

Hardt, M., Price, E., & Srebro, N. (2016). Equality of opportunity in supervised learning. *Advances in Neural Information Processing Systems, 29*, 3315–3323. https://proceedings.neurips.cc/paper/2016/hash/a486cd07e4ac3d270571622f4f316ec5-Abstract.html

Harris, M., & Johnson, C. (2020). Using AI to predict wildfires and mitigate their effects: A case study from California. *Journal of Climate Resilience, 7*(3), 215–228.

Hart, S. (1997). Beyond greening: Strategies for a sustainable world. *Harvard Business Review, 75*(1), 66–77.

Hassan, S. S., Meisner, K., Krause, K., Bzhalava, L., & Moog, P. (2024). Is digitalization a source of innovation? Exploring the role of digital diffusion in SME innovation performance. *Small Business Economics*, *62*(4), 1469–1491. DOI: 10.1007/s11187-023-00826-7

Heilinger, J.-C., Kempt, H., & Nagel, S. (2023). Beware of Sustainable AI! Uses and Abuses of a Worthy Goal. [Google Scholar] [CrossRef]. *AI and Ethics*, *4*, 1–12.

Henseler, J., Ringle, C. M., & Sarstedt, M. (2015). A new criterion for assessing discriminant validity in variance-based structural equation modeling. *Journal of the Academy of Marketing Science*, *43*(1), 115–135. DOI: 10.1007/s11747-014-0403-8

Hernández P., J., & Nishii, Y. (2021). Análisis de información y factores de desempeño ambiental y de economía circular en empresas peruanas. Comuni@cción, 12(1), 37-52. https://doi.org/DOI: 10.33595/2226-1478.12.1.481

Hinnell, A. C., Lazarovitch, N., Furman, A., Poulton, M., & Warrick, A. W. (2010). Neuro-drip: Estimation of subsurface wetting patterns for drip irrigation using neural networks. *Irrigation Science*, *28*(6), 535–544. DOI: 10.1007/s00271-010-0214-8

Hopkins, R. (2019). *From what is to what if: Unleashing the power of imagination to create the future we want.* Chelsea Green Publishing.

Hsieh, C.-J., Si, S., & Dhillon, I. A Divide-and-Conquer Solver for Kernel Support Vector Machines. In Proceedings of the 31st International Conference on Machine Learning, PMLR, Beijing, China, 22–24 June 2014; pp. 566–574.

Hu, Y. Y., Zhu, Y., Tucker, J., & Hu, Y. (2018). Ownership influence and CSR disclosure in China. *Accounting Research Journal*, *31*(1), 8–21. DOI: 10.1108/ARJ-01-2017-0011

Hwang, D., & Min, H. (2015). Identifying the drivers of enterprise resource planning and assessing its impacts on supply chain performances. *Industrial Management & Data Systems, 115*(3), 541–569. DOI: 10.1108/IMDS-10-2014-0284

Idrees, S. M., Alam, M. A., & Agarwal, P.Sheikh Mohammad Idrees. (2019). M Afshar Alam and Parul Agarwal, "A prediction approach for stock market volatility based on time series data". *IEEE Access : Practical Innovations, Open Solutions*, *7*, 17287–17298. DOI: 10.1109/ACCESS.2019.2895252

IEEE. (2019). *Ethically Aligned Design: A Vision for Prioritizing Human Well-being with Autonomous and Intelligent Systems.* https://standards.ieee.org/industry-connections/ec/ead.html

Iftikhar, A., Ali, I., Arslan, A., & Tarba, S. (2024). Digital innovation, data analytics, and supply chain resiliency: A bibliometric-based systematic literature review. *Annals of Operations Research, 333*(2), 825–848. DOI: 10.1007/s10479-022-04765-6 PMID: 35611176

Imalka, L. A., Gunawardana, K. G. A., Kodithuwakku, K. M. S. K., Arachchi, H. K. E., Harshanath, S. M. B., & Rajapaksha, S. (2022). Farming Through Technology Driven Solutions For Agriculture Industry Ceylon E-Agro mobile application-find technology based solutions for agricultural problems. In *2022 IEEE 10th Region 10 Humanitarian Technology Conference (R10-HTC).* DOI: 10.1109/R10-HTC54060.2022.9929335

Intergovernmental Panel on Climate Change (IPCC). (2022). Climate change 2022: Impacts, adaptation, and vulnerability. https://www.ipcc.ch/report/ar6/wg2/

IPCC. (2022). Climate Change 2022: Impacts, Adaptation and Vulnerability. Intergovernmental Panel on Climate Change. https://www.ipcc.ch/report/ar6/wg2/

Janiesch, C., Zschech, P., & Heinrich, K. (2021). Machine learning and deep learning. *Electronic Markets, 31*(3), 685–695. DOI: 10.1007/s12525-021-00475-2

Javaid, M., Haleem, A., Khan, I. H., & Suman, R. (2023). Understanding the potential applications of Artificial Intelligence in Agriculture Sector. *Advanced Agrochem, 2*(1), 15-30. DOI: 10.1016/j.aac.2022.10.001

Jha, K., Doshi, A., Patel, P., & Shah, M. (2019). A comprehensive review on automation in agriculture using artificial intelligence. *Artificial Intelligence in Agriculture*, *2*, 1–12. DOI: 10.1016/j.aiia.2019.05.004

Jiménez, A. P. (2024, April 30). *Artificial intelligence in the financial sector*. Retrieved from https://www.pwc.com/mx/es/liderazgo-estrategico/inteligencia-artificial-en-el-sector-financiero.html

Jobin, A., Ienca, M., & Vayena, E. (2019). The global landscape of AI ethics guidelines. *Nature Machine Intelligence, 1*(9), 389–399. DOI: 10.1038/s42256-019-0088-2

Jung, H. S., Lee, S. H., Lee, H., & Kim, J. H. (2023, August 1). Predicting Bitcoin Trends Through Machine Learning Using Sentiment Analysis with Technical Indicators. *Computer Systems Science and Engineering*, *46*(2), 2231–2246. DOI: 10.32604/csse.2023.034466

Kaack, L. H., Donti, P. L., Strubell, E., Kamiya, G., Creutzig, F., & Rolnick, D. (2022). Aligning artificial intelligence with climate change mitigation. *Nature Climate Change*, *12*(6), 518–527. DOI: 10.1038/s41558-022-01377-7

Kalyar, M. N., Shoukat, A., & Shafique, I. (2019). Enhancing firms' environmental performance and financial performance through green supply chain management practices and institutional pressures. Sustainability Accounting. *Management and Policy Journal*, *11*(2), 451–476. DOI: 10.1108/SAMPJ-02-2019-0047

Kamal, R. (2024). Cyborg Employees: Exploring the Benefits and Challenges of Transhumanism in Modern Work Environments. In *Embracing Transhumanism and Genomics in Human Resources Management* (pp. 1-20). IGI Global.

Kamal, R. (2024). Exploring the Synergy: Venture Capital's Impact on the Startup Ecosystem. In *Fostering Innovation in Venture Capital and Startup Ecosystems* (pp. 242-264). IGI Global.

Kamal, R., Kaur, M., Kaur, J., & Malhan, S. (2024). Artificial Intelligence-Powered Political Advertising: Harnessing Data-Driven Insights for Campaign Strategies. In *The Ethical Frontier of AI and Data Analysis* (pp. 100-109). IGI Global.

Kaur, J., Kamal, R., Kaur, M., & Malhan, S. (2024). The Symphony of Efficiency: Transforming Healthcare Through Robotic Process Automation. In *Clinical Practice and Unmet Challenges in AI-Enhanced Healthcare Systems* (pp. 245–261). IGI Global. DOI: 10.4018/979-8-3693-2703-6.ch013

Kaza, S., Yao, L. C., Bhada-Tata, P., & Van Woerden, F. (2018). *What a waste 2.0: A global snapshot of solid waste management to 2050.* World Bank Publications. DOI: 10.1596/978-1-4648-1329-0

Kemp, R., & Martens, P. (2007). Sustainable Development: How to Manage Something That Is Subjective and Never Can Be Achieved? *Sustainability, 3*(2), 5–14. DOI: 10.1080/15487733.2007.11907997

Kennedy, C., Cuddihy, J., & Engel-Yan, J. (2011). The study of urban metabolism and its applications to urban planning and design. *Environmental Pollution, 159*(8–9), 1965–1973. DOI: 10.1016/j.envpol.2011.02.023

Khanal, S., Kc, K., Fulton, J. P., Shearer, S., & Ozkan, E. (2020). Remote sensing in agriculture—Accomplishments, limitations, and opportunities. *Remote Sensing (Basel), 12*(22), 3783. DOI: 10.3390/rs12223783

Khan, S. A., & Qianli, D. (2017). Impact of green supply chain management practices on firms' performance: An empirical study from the perspective of Pakistan. *Environmental Science and Pollution Research International, 24*(20), 16829–16844. DOI: 10.1007/s11356-017-9172-5 PMID: 28573559

Khor, K. S., Thurasamy, R., Ahmad, N. H., Halim, H. A., & May-Chiun, L. (2015). Bridging the gap of green IT/IS and sustainable consumption. *Global Business Review, 16*(4), 571–593. DOI: 10.1177/0972150915581101

Kiani, F., Randazzo, G., Yelmen, I., Seyyedabbasi, A., Nematzadeh, S., Anka, F. A., Erenel, F., Zontul, M., Lanza, S., & Muzirafuti, A. (2022). A Smart and Mechanized Agricultural Application: From Cultivation to Harvest. *Applied Sciences (Basel, Switzerland), 12*(12), 6021. DOI: 10.3390/app12126021

Kim, Y., & Kaneda, M. (2018). Post-earthquake building resilience: Lessons from Japan's building code reform after the 1995 Kobe earthquake. *International Journal of Disaster Risk Reduction, 31*, 531–540. DOI: 10.1016/j.ijdrr.2018.07.004

Kotler, P., & Kartajaya, H. (n.d.). *Marketing 4.0.* LID.

Kurian, N., James, D., Varghese, V. S., Cherian, J. M., & Varghese, K. G. (2023). Artificial intelligence in scientific publications. *The Journal of the American Dental Association, 154*(12), 1041–1043. DOI: 10.1016/j.adaj.2023.03.015 PMID: 37140497

Kyriazis, D., Varvarigou, T., White, D., & Rossi, A. (2019). Sustainable smart city IoT applications: A novel eco-social model. *Internet of Things : Engineering Cyber Physical Human Systems, 1-2*, 4–10. DOI: 10.1016/j.iot.2018.08.001

Landrigan, P. J., Fuller, R., & Acosta, N. J. R.. (2018). The Lancet Commission on pollution and health. *Lancet, 391*(10119), 462–512. DOI: 10.1016/S0140-6736(17)32345-0

Lansing, J. S., & Kremer, J. N. (1993). Emergent Properties of Balinese Water Temple Networks: Coadaptation on a Rugged Fitness Landscape. *American Anthropologist, 95*(1), 97–114. DOI: 10.1525/aa.1993.95.1.02a00050

Lefebvre, H. (1968). Le droit à la ville. *Anthropos: Studie z Oboru Anthropologie, Paleoethnologie, Paleontologie a Kvartérní Geologie.*

Lei, J. (2024). Efficient Strategies on Supply Chain Network Optimization for Industrial Carbon Emission Reduction. *arXiv preprint arXiv:2404.16863*. Disponible en: https://arxiv.org/abs/2404.16863

Lei, C. F., Ngai, E. W., Lo, C. W., & See-To, E. W. (2023). Green IT/IS adoption and environmental performance: The synergistic roles of IT–business strategic alignment and environmental motivation. *Information & Management*, *60*(8), 103886. DOI: 10.1016/j.im.2023.103886

Lembcke, T. B., Herrenkind, B., Nastjuk, I., & Brendel, A. B. (2021). Promoting Business Trip Ridesharing with Green Information Systems: A Blended Environment Perspective. *Transportation Research Part D, Transport and Environment*, *94*, 102795. DOI: 10.1016/j.trd.2021.102795

Li, S., Jayaraman, V., Paulraj, A., & Shang, K. (2016). Proactive environmental strategies and performance: Role of green supply chain processes and green product design in the Chinese high-tech industry. *International Journal of Production Research*, *54*(7), 2136–2151. DOI: 10.1080/00207543.2015.1111532

Liu, F., Cai, M., Wang, L., & Lu, Y. (2019, February). An ensemble model based on adaptive noise reducer and over-fitting prevention LSTM for multivariate time series forecasting. *IEEE Access : Practical Innovations, Open Solutions*, *7*, 26102–26115. DOI: 10.1109/ACCESS.2019.2900371

Li, X., Li, J., & Wang, E. (2020). Sponge city construction in China: A policy perspective. *Environmental Science & Policy*, *109*, 14–22. DOI: 10.1016/j.envsci.2020.03.008

Li, Y., Pinto, M. C. B., & Diabat, A. (2020). Analyzing the critical success factor of CSR for the Chinese textile industry. *Journal of Cleaner Production*, *260*, 120878. DOI: 10.1016/j.jclepro.2020.120878

Lo, F. Y., & Liao, P. C. (2121). Rethinking financial performance and corporate sustainability: Perspectives on resources and strategies. Technological Forecasting and Social Change, 162, 120346. https://doi.org/ DOI: 10.1016/j.techfore.2020.120346

Lochan, K., Khan, A., Elsayed, I., Suthar, B., Seneviratne, L., & Hussain, I. (2024). Advancements in Precision Spraying of Agricultural Robots: A Comprehensive Review. *IEEE Access : Practical Innovations, Open Solutions*, *12*, 129447–129483. DOI: 10.1109/ACCESS.2024.3450904

Lodhi, S. K., Gill, A. Y., & Hussain, H. K. (2024). Green Innovations: Artificial Intelligence and Sustainable Materials in Production. *BULLET: Jurnal Multidisiplin Ilmu*, *3*(4), 492–507.

Loeser, F., Recker, J., vom Brocke, J., Molla, A., & Zarnekow, R. (2017). How IT executives create organizational benefits by translating environmental strategies into Green IS initiatives. *Information Systems Journal*, *27*(4), 503–553. DOI: 10.1111/isj.12136

Löser, F. (2015). Strategic Information Systems Management for Environmental Sustainability: Enhancing Firm Competitiveness with Green IS. Vol (6), Universitätsverlag der TU Berlin. https://doi.org/http://dx.doi .org/10.14279/depositonce-4447

Loy, D. (1985). Wei-Wu-Wei: Nondual Action. *Philosophy East & West*, *35*(1), 73–86. DOI: 10.2307/1398682

Lu, C. H. (2021). The impact of artificial intelligence on economic growth and welfare. *Journal of Macroeconomics*, *69*, 103342. DOI: 10.1016/j. jmacro.2021.103342

Luo, Q. Daoism and Environmental Sustainability—A Completely Different Way of Thinking. In *Proceedings of the International Workshop on Sustainable City Region*, Bali, Indonesia, 23–24 February 2009; pp. 164–171.

Luo, Y., Yin, L., Bai, W., & Mao, K. (2020). An Appraisal of Incremental Learning Methods. *Entropy (Basel, Switzerland)*, *22*(11), 1190. DOI: 10.3390/e22111190 PMID: 33286958

Mahlein, A.-K., Arnal Barbedo, J. G., Chiang, K.-S., Del Ponte, E. M., & Bock, C. H. (2024). From Detection to Protection: The Role of Optical Sensors, Robots, and Artificial Intelligence in Modern Plant Disease Management. *Phytopathology*, *114*(8), 1733–1741. DOI: 10.1094/PHYTO-01-24-0009-PER PMID: 38810274

Mahmud, M. S., Zahid, A., He, L., & Martin, P. (2021). Opportunities and possibilities of developing an advanced precision spraying system for tree fruits. *Sensors (Basel)*, *21*(9), 3262. DOI: 10.3390/s21093262 PMID: 34066785

Mahraz, M. I., Benabbou, L., & Berrado, A. (2020). A Compilation and Analysis of Critical Success Factors for the ERP Implementation. *International Journal of Enterprise Information Systems*, *16*(2), 107–133. DOI: 10.4018/IJEIS.2020040107

Marchese, A. J., McPhearson, T., & Pickett, S. T. A. (2018). Resilience and sustainability in urban systems. *Urban Ecosystems*, *21*(3), 423–438. DOI: 10.1007/s11252-018-0747-4

Martínez-Arroyo, J. A., Valenzo-Jiménez, M. A., & Kido-Cruz, A. (2022). Gestión ambiental, innovación y ventaja competitiva en empresas exportadoras de aguacate. *Investigación administrativa*, 51(129). https://doi.org/ DOI: 10.35426/IAv51n129.01

Masa'deh, R., Alananzeh, O., Algiatheen, N., Ryati, R., Albayyari, R., & Tarhini, A. (2017). The impact of employee's perception of implementing green supply chain management on hotel's economic and operational performance. *Journal of Hospitality and Tourism Technology*, *8*(3), 395–416. Advance online publication. DOI: 10.1108/JHTT-02-2017-0011

Meacham, J., Toms, L., Green, K.Jr, & Bhadauria, V. (2013). Impact of information sharing and green information systems. *Management Research Review*, *36*(5), 478–494. DOI: 10.1108/01409171311327244

Medellín Municipality. (2023). Social urbanism and the transformation of Medellín. https://www.medellin.gov.co/

Meenal, R., Binu, D., Ramya, K. C., Michael, P. A., Vinoth Kumar, K., Rajasekaran, E., & Sangeetha, B. (2022). Weather forecasting for renewable energy system: A review. *Archives of Computational Methods in Engineering*, *29*(5), 2875–2891. DOI: 10.1007/s11831-021-09695-3

Meerow, S., Newell, J. P., & Stults, M. (2016). Defining urban resilience: A review. *Landscape and Urban Planning*, *147*, 38–49. DOI: 10.1016/j.landurbplan.2015.11.011

Melville, N. (2010). Information Systems Innovation for Environmental Sustainability. *Management Information Systems Quarterly*, *34*(1), 1. DOI: 10.2307/20721412

Mensah, J. (2019). Mensah, J. Sustainable Development: Meaning, History, Principles, Pillars, and Implications for Human Action: Literature Review. *Cogent Social Sciences*, *5*(1), 1653531. DOI: 10.1080/23311886.2019.1653531

Mills, A. J., Durepos, G., & Wiebe, E. (Eds.). (2021). *Encyclopedia of Case Study Research.* SAGE Publications.

Mohamed, S., Png, M.-T., & Isaac, W. (2020). Decolonial AI: Decolonial theory as sociotechnical foresight in artificial intelligence. *Philosophy & Technology*, *33*(4), 659–684. DOI: 10.1007/s13347-020-00405-8

Molla, A., Cooper, V., & Pittayachawan, S. (2011). The Green IT Readiness (G-Readiness) of Organizations: An Exploratory Analysis of a Construct and Instrument. *Communications of the Association for Information Systems*, *29*(4), 67–96. https://aisel.aisnet.org/cais/vol29/iss1/4. DOI: 10.17705/1CAIS.02904

Morley, J., Cowls, J., Taddeo, M., & Floridi, L. (2020). Ethical guidelines for COVID-19 tracing apps. *Nature*, *582*(7810), 29–31. DOI: 10.1038/d41586-020-01578-0 PMID: 32467596

Mueller, N., Rojas-Rueda, D., & Tainio, M. (2015). Health impact assessment of active transport: Benefits of cycling and walking in urban environments. *Environmental Health Perspectives*, *123*(7), 640–647. DOI: 10.1289/ehp.1409091

Nagler, J., van den Hoven, J., & Helbing, D. (2019). An Extension of Asimov's Robotics Laws. In Helbing, D. (Ed.), *Towards Digital Enlightenment*. Springer. DOI: 10.1007/978-3-319-90869-4_5

Nahar, S. (2024). Modeling the effects of artificial intelligence (AI)-based innovation on sustainable development goals (SDGs): Applying a system dynamics perspective in a cross-country setting. *Technological Forecasting and Social Change*, *201*, 123203. DOI: 10.1016/j.techfore.2023.123203

Nanath, K., & Pillai, R. R. (2017). The Influence of Green IS Practices on Competitive Advantage: Mediation Role of Green Innovation Performance. *Information Systems Management*, *34*(1), 3–19. DOI: 10.1080/10580530.2017.1254436

Nash, K. L., Blythe, J. L., Cvitanovic, C., Fulton, E. A., Halpern, B. S., Milner-Gulland, E., Addison, P. F., Pecl, G. T., Watson, R. A., & Blanchard, J. L. (2020). To Achieve a Sustainable Blue Future, Progress Assessments Must Include Interdependencies between the Sustainable Development Goals. *One Earth*, *2*(2), 161–173. DOI: 10.1016/j.oneear.2020.01.008

Nelson, E. S. (2014). Technology and the Way: Buber, Heidegger, and Lao-Zhuang "Daoism". *Journal of Chinese Philosophy*, *41*(3-4), 307–327. DOI: 10.1163/15406253-0410304005

Nelson, E. S. (2020). *Daoism and Environmental Philosophy: Nourishing Life*. Routledge. DOI: 10.4324/9780429399145

Newman, P., & Kenworthy, J. (2021). *The end of automobile dependence: How cities are moving beyond car-based planning*. Island Press.

Nguyen, A.-P., Moreno, D. L., Le-Bel, N., & Rodríguez Martínez, M. (2023). MonoNet: Enhancing Interpretability in Neural Networks via Monotonic Features. *Bioinformatics Advances*, *3*(1), vbad016. DOI: 10.1093/bioadv/vbad016 PMID: 37143924

Nicolás, J. J. (2021, May 18). *Getting to know A9: Dissecting Amazon's algorithm*. Retrieved from https://es.linkedin.com/pulse/conociendo-a9-destripando-el-algoritmode-amazon-jose-javier-nicol%C3%A1s

Nordgren, A. (2023). Artificial Intelligence and Climate Change: Ethical Issues. *J. Inf. Commun. Ethics Soc.*, *21*(1), 1–15. DOI: 10.1108/JICES-11-2021-0106

Norheim, H. (2018). The virtues of green strategies: Some empirical support from the Alliance context. *Journal of Business Ethics, 151*(4), 1161–1173. DOI: 10.1007/s10551-016-3274-8

Nyéki, A., & Neményi, M. (2022). Crop Yield Prediction in Precision Agriculture. *Agronomy (Basel), 12*(10), 2460. DOI: 10.3390/agronomy12102460

OECD. (2023c). *Governing cities: Bringing together urban and regional development.* OECD Publishing., DOI: 10.1787/9789264291943-

Ofek, N., Katz, G., Shapira, B., & Bar-Zev, Y. Sentiment Analysis in Transcribed Utterances. In Advances in Knowledge Discovery and Data Mining; Lecture Notes in Computer Science; 2020.

Ofek, N., Rokach, L., Stern, R., & Shabtai, A. (2017). Fast-CBUS: A Fast Clustering-Based Undersampling Method for Addressing the Class Imbalance Problem. *Neurocomputing, 243*, 88–102. DOI: 10.1016/j.neucom.2017.03.011

Ok, A. O., Akar, O., & Gungor, O. (2012). Evaluation of random forest method for agricultural crop classification. *European Journal of Remote Sensing, 45*(1), 421–432. DOI: 10.5721/EuJRS20124535

Olsson, P., Galaz, V., & Boonstra, W. J. (2015). Sustainability transformations: A resilience perspective. *Ecology and Society, 20*(3), 1–19. DOI: 10.5751/ES-07887-200341

Open Web Application Security Project (OWASP). (n.d.). *OWASP*. Retrieved from https://owasp.org/

Organisation for Economic Co-operation and Development (OECD). (2023). Affordable housing and urban inequality: OECD statistics. https://www.oecd.org/urban/affordable-housing

Organisation for Economic Co-operation and Development (OECD). (2023). Urban governance and the challenges of fragmentation: OECD report. https://www.oecd.org

Organisation for Economic Co-operation and Development. (2023a). Cities policy responses to COVID-19: OECD report. OECD. https://www.oecd.org/coronavirus/policy-responses/cities-policy-responses-to-covid-19-df3f6e1d/

Pagano, T. P., Loureiro, R. B., Lisboa, F. V. N., Peixoto, R. M., Guimarães, G. A. S., Cruz, G. O. R., Araujo, M. M., Santos, L. L., Cruz, M. A. S., Oliveira, E. L. S., Winkler, I., & Nascimento, E. G. S. (2023). Bias and Unfairness in Machine Learning Models: A Systematic Review on Datasets, Tools, Fairness Metrics, and Identification and Mitigation Methods. *Big Data and Cognitive Computing*, *7*(1), 15. DOI: 10.3390/bdcc7010015

Pajares Mrtinsanz, G., & García, S. (2006). *Artificial Intelligence and Knowledge Engineering*. Alfaomega.

Pandey, P., Mishra, S., Rai, P., & Anand, A. (2019). Social engineering and exploit development. [IJSRCSAMS]. *International Journal of Scientific Research in Computer Science Applications and Management Studies*, *8*(5).

Pandian, P., Bose, P., Uma, M., Susithra, K., Panchami, B. G., & Reshma, R. (2024). Intelligent Sensor Based Smart Irrigation System For Sustainable Agriculture. *2024 International Conference on Communication, Computing and Internet of Things (IC3IoT)*. DOI: 10.1109/IC3IoT60841.2024.10550413

Pano, T., & Kashef, R. (2020). A Corpus of BTC Tweets in the Era of COVID-19. 2020 IEEE International IOT, Electronics and Mechatronics Conference (IEMTRONICS) (pp. 1-4), IEEE. DOI: 10.1109/IEMTRONICS51293.2020.9216427

Panpatte, D. G. (2018). Artificial intelligence in agriculture: An emerging era of research. Anand Agricultural University, 1-8.

Parolin, G., Mcaloone, T. C., & Pigosso, D. C. (2024). How can technology assessment tools support sustainable innovation? A systematic literature review and synthesis. *Technovation*, *129*, 102881. DOI: 10.1016/j.technovation.2023.102881

Pasquale, F. (2015). *The black box society: The secret algorithms that control money and information*. Harvard University Press. DOI: 10.4159/harvard.9780674736061

Patton, M. Q. (2015). *Qualitative research & evaluation methods: Integrating theory and practice* (4th ed.). SAGE Publications.

Pauleit, S., Zölch, T., Hansen, R., Randrup, T. B., & Konijnendijk van den Bosch, C. (2017). Nature-based solutions and climate change–four shades of green. Nature-based solutions to climate change adaptation in urban areas: Linkages between science, policy and practice, 29-49.

Pérez, L. M. (2021). *Factores críticos de éxito y fracaso de la transformación digital en las empresas. Tesis de maestría, Universidad de Talca.* https://dspace.utalca.cl/bitstream/1950/13204/3/2021A001044.pdf

Petter, S., DeLone, W. H., & McLean, E. R. (2013). Information Systems Success: The Quest for the Independent Variables. *Journal of Management Information Systems*, *29*(4), 7–62. DOI: 10.2753/MIS0742-1222290401

Pierce, J., & Jameton, A. (2004). *The Ethics of Environmentally Responsible Health Care*. Oxford University Press.

Pierre, N., Viviane, I. V. I., Lambert, U., Shadrack, I., Erneste, B., Schadrack, N., & Theogene, H. (2023). AI Based Real-Time Weather Condition Prediction with Optimized Agricultural Resources. *European Journal of Technology*, *7*(2), 36–49. DOI: 10.47672/ejt.1496

Pooja, P. (2024). Environmental Degradation: Causes, Impacts and Mitigation. *Journal of Advances in Science and Technology*, *20*(1), 220–223. DOI: 10.29070/94vmdz61

Portier, K., Greer, G. E., Rokach, L., Ofek, N., Wang, Y., Biyani, P., Yu, M., Banerjee, S., Zhao, K., Mitra, P., & Yen, J. (2013). Understanding Topics and Sentiment in an Online Cancer Survivor Community. *Journal of the National Cancer Institute. Monographs*, *2013*(47), 195–198. DOI: 10.1093/jncimonographs/lgt025 PMID: 24395991

Premkumar, S., & Sigappi, A. N. (2022). IoT-enabled edge computing model for smart irrigation system. *Journal of Intelligent Systems*, *31*(1), 632–650. DOI: 10.1515/jisys-2022-0046

Provost, F., & Kolluri, V. (1999). A Survey of Methods for Scaling Up Inductive Algorithms. *Data Mining and Knowledge Discovery*, *3*(2), 131–169. DOI: 10.1023/A:1009876119989

Purnomo, A., Susanti, M., & Widodo, W. (2022). *A study of green information technology using bibliometric analysis*. Springer Nature Singapore Pte Ltd., DOI: 10.1007/978-981-16-9573-5_46

Qazi, S., Khawaja, B. A., & Farooq, Q. U. (2022). IoT-Equipped and AI-Enabled Next Generation Smart Agriculture: A Critical Review, Current Challenges and Future Trends. *IEEE Access : Practical Innovations, Open Solutions*, *10*, 21219–21235. DOI: 10.1109/ACCESS.2022.3152544

Qin, Y., Xu, Z., Wang, X., & Skare, M. (2024). Artificial intelligence and economic development: An evolutionary investigation and systematic review. *Journal of the Knowledge Economy*, *15*(1), 1736–1770. DOI: 10.1007/s13132-023-01183-2

Quiroz, D., & Vera, C. (2023). *Ética ambiental corporativa e innovación verde: Un modelo de gestión empresarial. Revista Facultad de Ciencias Económicas, 31*(1), 75-92. https://revistas.unimilitar.edu.co/index.php/rfce/article/view/6715/5862

Qu, T., & Liu, Y. (2022). Green innovations, supply chain integration and green information system: A model of moderation. *Journal of Cleaner Production*, *340*, 130667. DOI: 10.1016/j.jclepro.2022.130557

Rahwan, I. (2018). Society-in-the-loop: Programming the algorithmic social contract. *Ethics and Information Technology*, *20*(1), 5–14. DOI: 10.1007/s10676-017-9430-8

Rajan, C. A., & Baral, R. (2015). Adoption of ERP system: An empirical study of factors influencing the usage of ERP and its impact on end user. *IIMB Management Review*, *27*(2), 105–117. DOI: 10.1016/j.iimb.2015.04.008

Ramani, K., Jahnavi, M., Reddy, P. J., VenkataChakravarthi, P., Meghanath, P., & Imran, S. K. (2023). Prediction of bitcoin price through LSTM, ARIMA, XGBoost, prophet and sentiment analysis on dynamic streaming data. 2023 9th International Conference on Advanced Computing and Communication Systems (ICACCS) (Vol. 1, pp. 1514-1518), IEEE.

Ramaswami, A., Hillman, T., Janson, B., Reiner, M., & Thomas, G. (2017). A social–ecological approach to the urban metabolism of cities. *Nature Sustainability, 1*(5), 213–222. DOI: 10.1038/s41893-018-0033-3

Ram, J., & Corkindale, D. (2014). How "critical" are the critical success factors (CSFs)? Examining the role of CSFs for ERP. *Business Process Management Journal, 20*(1), 151–174. DOI: 10.1108/BPMJ-11-2012-0127

Rani, R., Sahoo, J., Bellamkonda, S., Kumar, S., & Pippal, S. K. (2023). Role of Artificial Intelligence in Agriculture: An Analysis and Advancements With Focus on Plant Diseases. *IEEE Access : Practical Innovations, Open Solutions, 11*, 137999–138019. DOI: 10.1109/ACCESS.2023.3339375

Raworth, K. (2017). *Doughnut economics: Seven ways to think like a 21st-century economist*. Chelsea Green Publishing.

Reddy, R., Sirish Kumar, M., Shereesha, M., Raju, K. S., Tangudu, N., & Nagendra Babu, G. (2023). A Comprehensive Overview on Implementing IoT-AI Interface in Agriculture. *2023 Second International Conference on Augmented Intelligence and Sustainable Systems (ICAISS)*. DOI: 10.1109/ICAISS58487.2023.10250697

Reitsma, E., & Hilletofth, P. (2018). Critical success factors for ERP system implementation: A user perspective. *European Business Review, 30*(3), 285–310. DOI: 10.1108/EBR-04-2017-0075

Report: Inequalities Exacerbate Climate Impacts on Poor. Available online: https://www.un.org/sustainabledevelopment/blog/2016/10/report-inequalities-exacerbate-climate-impacts-on-poor/(accessed on 24 September 2023).

Resilient Cities Network. (2023). Urban resilience for whom, what, when, where, and why? https://www.resilientcitiesnetwork.org/urban-resilience

Rolnick, D., Donti, P. L., Kaack, L. H., Kochanski, K., Lacoste, A., Sankaran, K., . . . Bengio, Y. (2019). Tackling climate change with machine learning. *arXiv preprint arXiv:1906.05433*. https://arxiv.org/abs/1906.05433

Rolnik, R., Giraudo, A., & Cardoso, A. (2021). Minha Casa Minha Vida: The challenges of large-scale affordable housing in Brazil. *Housing Studies*, *36*(1), 123–145. DOI: 10.1080/02673037.2020.1789642

Rose, S., Borchert, O., Mitchell, S., & Connelly, S. (2020). *Zero trust architecture* (NIST Special Publication 800-207). National Institute of Standards and Technology. DOI: 10.6028/NIST.SP.800-207

Rotz, S., Gravely, E., Mosby, I., Duncan, E., Finnis, E., Horgan, M., LeBlanc, J., Martin, R., Neufeld, H. T., Nixon, A., Pant, L., Shalla, V., & Fraser, E. D. G. (2019). Automated pastures and the digital divide: How agricultural technologies are shaping labour and rural communities. *Journal of Rural Studies*, *68*, 112–122. DOI: 10.1016/j.jrurstud.2019.01.023

Rouhiainen, J. (2019). *Artificial intelligence for business*. Retrieved June 11, 2024, from https://repositorio.usil.edu.pe/server/api/core/bitstreams/e190d249-c1ca-4006-bf5f-cf1b0d0a12cd/content

Ruiz, E. I. M., Almazán, D. A., & Cortés, C. B. Y. (2023). Multidimensional scale for assessing the determinants of success of green information systems on organizational performance. *Revista del Centro de Investigación de la Universidad la Salle*, *15*(59), 105–130. DOI: 10.26457/recein.v15i59.3524

Rutherford, J. (2014). The Vicissitudes of District Energy: Consolidation, loss and disconnection in Stockholm's district heating grid. *Urban Studies (Edinburgh, Scotland)*, *51*(5), 1055–1071. DOI: 10.1177/0042098013495573

Rutherford, L. (2014). Decentralized governance and sustainable urban development in Sweden and Switzerland. *Sustainable Cities and Society*, *12*, 38–49. DOI: 10.1016/j.scs.2014.03.001

Ryoo, S. Y., & Koo, C. (2013). Green practices-IS alignment and environmental performance: The mediating effects of coordination. *Information Systems Frontiers*, *15*(5), 799–814. DOI: 10.1007/s10796-013-9422-0

Saade, R. G., & Nijher, H. (2016). Critical success factors in enterprise resource planning implementation: A review of case studies. *Journal of Enterprise Information Management*, *29*(1), 72–96. DOI: 10.1108/JEIM-03-2014-0028

Sachs, J. D., Schmidt-Traub, G., Kroll, C., Lafortune, G., & Fuller, G. (2021). *Sustainable development report 2021: The Decade of Action for the Sustainable Development Goals*. Cambridge University Press.

Salaheldeen, M., & Battour, M. (2024). Fostering innovation capability and sustainable innovation in halal industry: The role of halal entrepreneurs' success. *Journal of Islamic Marketing*, *15*(3), 777–799. DOI: 10.1108/JIMA-12-2022-0323

Saleem, T., Yaqub, U., & Zaman, S. (2024, February 19). Twitter sentiment analysis and bitcoin price forecasting: Implications for financial risk management. *The Journal of Risk Finance*, *25*(3), 407–421. DOI: 10.1108/JRF-09-2023-0241

Salvia, A. L., Leal Filho, W., Brandli, L. L., & Griebeler, J. S. (2019). Assessing research trends related to sustainable development goals: Local and global issues. *Journal of Cleaner Production*, *208*, 841–849. DOI: 10.1016/j.jclepro.2018.09.242

Samaniego, M., & Deters, R. (2018). Zero-trust hierarchical management in IoT. In *Proceedings of the IEEE International Congress on Internet of Things (ICIOT)* (pp. 88–95).

Sánchez Balcázar, M. C., & Gómez, E. (2023). *Effects of the application of artificial intelligence in accounting and decision-making*. https://revistap.ejeutap.edu.co/index.php/Gestion/article/view/71

Sarkar, P., & Young, L. (2009). Managerial attitudes towards Green IT: An explorative study of policy drivers. PACIS 2009 Proceedings, 1–14. https://aisel.aisnet.org/cgi/viewcontent.cgi?article=1071&context=pacis2009

Sassen, S. (2018). *Expulsions: Brutality and complexity in the global economy*. Belknap Press.

Sawyer, R. K., & Henriksen, D. (2024). *Explaining creativity: The science of human innovation*. Oxford university press.

Saxena, P. K., Seetharaman, A., & Shawarikar, G. (2024). Factors That Influence Sustainable Innovation in Organizations: A Systematic Literature Review. *Sustainability (Basel)*, *16*(12), 4978. DOI: 10.3390/su16124978

Saygili, E. E., Ozturkoglu, Y., & Kocakulah, M. C. (2017). End users' perceptions of critical success factors in ERP applications. *International Journal of Enterprise Information Systems*, *13*(4), 58–75. DOI: 10.4018/IJEIS.2017100104

Schwartz, R., Dodge, J., Smith, N., & Etzioni, O. (2020). Green AI. *Communications of the ACM*, *63*(12), 54–63. DOI: 10.1145/3381831

Seeley, T. D. (2002). When Is Self-Organization Used in Biological Systems? *The Biological Bulletin*, *202*(3), 314–318. DOI: 10.2307/1543484 PMID: 12087005

Seidel, S., Bharati, P., Watson, R. T., & Boudreau, M. (2017). The Sustainability Imperative in Information Systems Research. *Communications of the Association for Information Systems*, *40*(1), 40–52. Advance online publication. DOI: 10.17705/1CAIS.04003

Sengupta, S., Choudhary, S., Obayi, R., & Nayak, R. (2024). Reducing food loss through sustainable business models and agricultural innovation systems. *Supply Chain Management*, *29*(3), 540–572. DOI: 10.1108/SCM-01-2023-0059

Sepasgozar, S. M. (2021). Differentiating digital twin from digital shadow: Elucidating a paradigm shift to expedite a smart, sustainable built environment. *Buildings (Basel, Switzerland)*, *11*(4), 151. DOI: 10.3390/buildings11040151

Shah, N., Zehri, A. W., Saraih, U. N., Abdelwahed, N. A. A., & Soomro, B. A. (2024). The role of digital technology and digital innovation towards firm performance in a digital economy. *Kybernetes*, *53*(2), 620–644. DOI: 10.1108/K-01-2023-0124

Shaikh, F. K., Memon, M. A., Mahoto, N. A., Zeadally, S., & Nebhen, J. (2022). Artificial Intelligence Best Practices in Smart Agriculture. *IEEE Micro*, 42(1), 17-24. .DOI: 10.1109/MM.2021.3121279

Shin, D.-H., & Park, Y. J. (2019). Role of fairness, accountability, and transparency in algorithmic affordance. *Computers in Human Behavior*, *98*, 277–284. DOI: 10.1016/j.chb.2019.04.019

Simeon, I. A., Caballero, R., & Lepez, C. O. (2024, February). Curricular innovation in the knowledge society. In *Seminars in Medical Writing and Education* (Vol. 3, pp. 58-58).

Singapore Government Technology Agency. (2023). AI-based waste management in Singapore: Optimizing urban sustainability. https://www.tech.gov.sg

Singapore Urban Redevelopment Authority. (2023). Smart and sustainable urban design in Singapore. https://www.ura.gov.sg

Singapore Urban Redevelopment Authority. (2023). Urban sustainability and smart nation initiatives. https://www.ura.gov.sg/

Singapore, P. U. B. (2023). Singapore's water story: Four national taps. https://www.pub.gov.sg/

Singapore, P. U. B. (2023). Water management and sustainability in Singapore: The Four National Taps policy. https://www.pub.gov.sg

Singh, V., & Singh, G. (2023). Development of AI Enabled Solution for Efficient Feature Enrichment from Multiple Data Sources: An Application in Precision Agriculture. *2023 International Conference on Artificial Intelligence and Applications (ICAIA) Alliance Technology Conference (ATCON-1)*. DOI: 10.1109/ICAIA57370.2023.10169451

Singh, A., Kanaujia, A., Singh, V. K., & Vinuesa, R. (2024). Artificial intelligence for Sustainable Development Goals: Bibliometric patterns and concept evolution trajectories. *Sustainable Development (Bradford), 32*(1), 724–754. DOI: 10.1002/sd.2706

Singh, P. K., & Maheswaran, R. (2024). Analysis of social barriers to sustainable innovation and digitisation in supply chain. *Environment, Development and Sustainability, 26*(2), 5223–5248. DOI: 10.1007/s10668-023-02931-9 PMID: 36687733

Sinwar, D., Dhaka, V. S., Sharma, M. K., & Rani, G. (2020). AI-based yield prediction and smart irrigation. *Internet of Things and Analytics for Agriculture, 2*, 155–180.

Sison, A. J. G., Daza, M. T., Gozalo-Brizuela, R., & Garrido-Merchán, E. C. ChatGPT: More than a Weapon of Mass Deception, Ethical Challenges and Responses from the Human-Centered Artificial Intelligence (HCAI) Perspective. arXiv 2023, arXiv:2304.11215. [Google Scholar]

Smart Nation Singapore. (2023). The Smart Nation initiative: Using technology for sustainable urban development. https://www.smartnation.gov.sg

Smart Nation Singapore. (2023). What is Smart Nation? https://www.smartnation.gov.sg/

Soja, E. W. (2010). *Seeking spatial justice*. University of Minnesota Press.

Srivastava, P. K., Han, D., Rico-Ramirez, M. A., Bray, M., & Islam, T. (2012). Selection of classification techniques for land use/land cover change investigation. *Advances in Space Research, 50*(9), 1250–1265. DOI: 10.1016/j.asr.2012.06.032

State of Green. (2023). Copenhagen's renewable energy initiatives and district heating system. https://stateofgreen.com/en/partners/copenhagen

State of Green. (2023). Copenhagen's smart energy system: A sustainable model for the future. https://stateofgreen.com/en/partners/copenhagen

State of Green. (2023). Denmark's green transition: Leading by example. https://stateofgreen.com/

Statista. (n.d.). *Revenue from OTT and video across India from 2018 to 2022, with estimates until 2027*. Retrieved from https://www.statista.com/statistics/1231740/ott-and-video-revenue-india/

Steinhoff, J. (2024). AI ethics as subordinated innovation network. *AI & Society*, *39*(4), 1995–2007. DOI: 10.1007/s00146-023-01658-5

Stickler, L. (2022). The economic costs of urban traffic congestion. *Urban Economics Journal*, *12*(4), 456–470.

Strubell, E., Ganesh, A., & McCallum, A. (2019). Energy and policy considerations for deep learning in NLP. *Proceedings of the 57th Annual Meeting of the Association for Computational Linguistics*, 3645–3650. DOI: 10.18653/v1/P19-1355

Subeesh, A., & Mehta, C. R. (2021). Automation and digitization of agriculture using artificial intelligence and internet of things. *Artificial Intelligence in Agriculture*, *5*, 278–291. DOI: 10.1016/j.aiia.2021.11.004

Sun, Y., Song, H., Jara, A. J., & Bie, R. (2021). Internet of Things and big data analytics for smart and connected communities. *Journal of Environmental Research and Public Health*, *18*(9), 4784. DOI: 10.3390/ijerph18094784 PMID: 33946141

Suša Vugec, D., Bosilj Vukšić, V., Pejić Bach, M., Jaklič, J., & Indihar Štemberger, M. (2020). Business intelligence and organizational performance: The role of alignment with business process management. *Business Process Management Journal*, *26*(6), 1709–1730. DOI: 10.1108/BPMJ-08-2019-0342

Sustainable Development Solutions Network. (2019). *The role of technology in delivering the sustainable development goals*. United Nations.

Talib, M. S., Hamid, A. B. A., & Thoo, A. C. (2015). Critical success factors of supply chain management: A literature survey and Pareto analysis. *EuroMed Journal of Business*, *10*(2), 234–263. DOI: 10.1108/EMJB-09-2014-0028

Tesla. (2023). Energy products: Powerpack and Powerwall. https://www.tesla.com/energy

Topol, E. J. (2019). *Deep medicine: How artificial intelligence can make healthcare human again*. Basic Books.

Trenerry, B., Chng, S., Wang, Y., Suhaila, Z. S., Lim, S. S., Lu, H. Y., & Oh, P. H. (2021). Preparing workplaces for digital transformation: An integrative review and framework of multi-level factors. *Frontiers in Psychology*, *12*, 620766. DOI: 10.3389/fpsyg.2021.620766 PMID: 33833714

Tripathy, N., Hota, S., & Mishra, D., Satapathy, P., & Nayak, S. K. (2024). Empirical Forecasting Analysis of Bitcoin Prices: A Comparison of Machine learning, Deep learning, and Ensemble learning Models. *International journal of electrical and computer engineering systems, 15*(1), 21-9. doi.org/10.32985/ijeces.15.1.3

Tripathy, N., Balabantaray, S. K., Parida, S., & Nayak, S. K. (2024, June). Cryptocurrency fraud detection through classification techniques [IJECE]. *International Journal of Electrical and Computer Engineering*, *14*(3), 2918–2926. DOI: 10.11591/ijece.v14i3.pp2918-2926

Tripathy, N., Hota, S., & Mishra, D. (2023, September). Performance analysis of bitcoin forecasting using deep learning techniques. *Indonesian Journal of Electrical Engineering and Computer Science*, *31*(3), 1515–1522. DOI: 10.11591/ijeecs.v31.i3.pp1515-1522

Tripathy, N., Hota, S., Prusty, S., & Nayak, S. K. (2023). Performance Analysis of Deep Learning Techniques for Time Series Forecasting", In*2023 International Conference in Advances in Power, Signal, and Information Technology (APSIT)* (pp. 639-644). IEEE. DOI: 10.1109/APSIT58554.2023.10201734

Tummers, L. (2016). Innovative housing strategies for sustainable cities: Adaptive reuse, incremental housing, and co-housing. *Sustainable Development*, *24*(6), 326–340. DOI: 10.1002/sd.1602

Ulfa, F., Orton, T. G., Dang, Y. P., & Menzies, N. W. (2022). Developing and Testing Remote-Sensing Indices to Represent within-Field Variation of Wheat Yields: Assessment of the Variation Explained by Simple Models. *Agronomy (Basel)*, *12*(2), 384. DOI: 10.3390/agronomy12020384

UNESCO. (2021). *Recommendation on the Ethics of Artificial Intelligence*. https://unesdoc.unesco.org

UNESCO. (2021). *Recommendation on the Ethics of Artificial Intelligence*. United Nations Educational, Scientific and Cultural Organization.

UNESCO. (2022). Culture and sustainable development. https://en.unesco.org/themes/culture-and-sustainable-development

UN-Habitat. (2020). World Cities Report 2020: The value of sustainable urbanization. United Nations Human Settlements Programme. https://unhabitat.org/wcr/

UN-Habitat. (2023). *World cities report 2022: Envisaging the future of cities*. United Nations Human Settlements Programme.

UN-Habitat. (2023). World cities report 2023: The value of sustainable urbanization. https://unhabitat.org/world-cities-report-2023

United Nations Department of Economic and Social Affairs. (2023). *Goal 11: Sustainable cities and communities*. United Nations.

United Nations Department of Economic and Social Affairs. (2023). *World population prospects 2022: Summary of results*. United Nations.

United Nations Department of Economic and Social Affairs. (2023). World population prospects 2022: Summary of results. United Nations. https://population.un.org/wpp/Publications/Files/WPP2022_Summary-of-Results.pdf

United Nations Department of Economic and Social Affairs. (2023). World urbanization prospects 2022: Highlights. United Nations. https://population.un.org/wup/Publications/Files/WUP2022_Highlights.pdf

United Nations Human Settlements Programme (UN-Habitat). (2020). Urban governance for sustainable development. https://unhabitat.org

United Nations Office for Disaster Risk Reduction (UNDRR). (2015). Sendai Framework for Disaster Risk Reduction 2015–2030. https://www.undrr.org/implementing-sendai-framework

United Nations. (2015). *Transforming our world: The 2030 agenda for sustainable development.*

Vaishya, R., Javaid, M., Khan, I. H., & Haleem, A. (2020). Artificial intelligence (AI) applications for COVID-19 pandemic. *Diabetes & Metabolic Syndrome*, *14*(4), 337–339. DOI: 10.1016/j.dsx.2020.04.012 PMID: 32305024

Van Wynsberghe, A., & Sustainable, A. I. (2021). AI for Sustainability and the Sustainability of AI. *AI and Ethics*, *1*(3), 213–218. DOI: 10.1007/s43681-021-00043-6

Vargo, S. L., Fehrer, J. A., Wieland, H., & Nariswari, A. (2024). The nature and fundamental elements of digital service innovation. *Journal of Service Management*, *35*(2), 227–252. DOI: 10.1108/JOSM-02-2023-0052

Velásquez, T., Messino, A., Navarro, H. E., & Ovalle, A. A. (2019). The role of Green IT in the non-functional software development requirements: Perspectives for functional design. *Journal of Physics: Conference Series*, *1388*(1), 012030. DOI: 10.1088/1742-6596/1388/1/012030

Vinuesa, R., Azizpour, H., Leite, I., Balaam, M., Dignum, V., Domisch, S., & Nerini, F. F. (2020). The role of artificial intelligence in achieving the Sustainable Development Goals. *Nature Communications*, *11*(1), 233. DOI: 10.1038/s41467-019-14108-y PMID: 31932590

Waltersmann, L., Kiemel, S., Stuhlsatz, J., Sauer, A., & Miehe, R. (2021). Artificial intelligence applications for increasing resource efficiency in manufacturing companies—A comprehensive review. *Sustainability (Basel)*, *13*(12), 6689. DOI: 10.3390/su13126689

Wamba, S. F., & Queiroz, M. M. (2020). Blockchain in the operations and supply chain management: Benefits, challenges and future research opportunities. *International Journal of Information Management, 52*, 102064. DOI: 10.1016/j.ijinfomgt.2019.102064

Wang, D., He, H., & Liu, D. (2019, July). Intelligent optimal control with critic learning for a nonlinear overhead crane system. *IEEE Transactions on Industrial Informatics, 14*(7), 2932–2940. DOI: 10.1109/TII.2017.2771256

Wang, H. (2024). Linking AI supply chain strength to sustainable development and innovation: A country-level analysis. *Expert Systems: International Journal of Knowledge Engineering and Neural Networks, 41*(5), e12973. DOI: 10.1111/exsy.12973

Wang, J., Cao, J., Yuan, S., & Cheng, M. (2021). Short-term forecasting of natural gas prices by using a novel hybrid method based on a combination of the CEEMDAN-SE-and the PSO-ALS-optimized GRU network. *Energy, 233*(121082), 121082. DOI: 10.1016/j.energy.2021.121082

Watson, R. T., Boudreau, M., & Chen, A. J. (2010). Information Systems and Environmentally Sustainable Development: Energy Informatics and New Directions for the IS Community. *Management Information Systems Quarterly, 34*(1), 23. DOI: 10.2307/20721413

Weiss, K., Khoshgoftaar, T. M., & Wang, D. (2016). A Survey of Transfer Learning. *Journal of Big Data, 3*(1), 9. DOI: 10.1186/s40537-016-0043-6

Weng, T., Liu, W., & Xiao, J. (2019, September). W. Liu and and J. Xiao, "Supply chain sales forecasting based on lightGBM and LSTM combination model". *Industrial Management & Data Systems, 120*(2), 265–279. DOI: 10.1108/IMDS-03-2019-0170

Wirth, P., Garschagen, M., & Cloke, H. (2022). Urban resilience transformation in Medellín: Insights from twenty years of social urbanism. *Sustainability, 14*(2), 891. DOI: 10.3390/su14020891

Wolman, A. (1965). The metabolism of cities. *Scientific American, 213*(3), 179–190. DOI: 10.1038/scientificamerican0965-179

World Bank. (2022). The world bank annual report 2022: Financing development through a green, resilient, and inclusive recovery. World Bank Group. https://www.worldbank.org/en/publication/annual-report

World Bank. (2023). Financing sustainable cities: Challenges and opportunities for SDG 11. https://www.worldbank.org/en/topic/urbandevelopment

World Business Council for Sustainable Development. (2023). Business solutions for sustainable cities. https://www.wbcsd.org/

World Health Organization (WHO). (2023). Air pollution and health. https://www.who.int/health-topics/air-pollution

World Health Organization. (2023). Air pollution levels rising in many of the world's poorest cities. https://www.who.int/news/item/29-09-2023-air-pollution-levels-rising-in-many-of-the-world's-poorest-cities

Wu, Q., Wu, C., Yan, X., & Cheng, Q. (2021). Intrinsic security and self-adaptive cooperative protection enabling cloud-native network slicing. *IEEE Transactions on Network and Service Management*, *18*(2), 1834–1848.

Wu, S., & Wu, Y. (2014). The influence of enterprisers' green management awareness on green management strategy and organizational performance. *International Journal of Quality & Reliability Management*, *31*(4), 455–476. DOI: 10.1108/IJQRM-01-2013-0019

Wu, X., Gomes-Selman, J., Shi, Q., Xue, Y., Garcia-Villacorta, R., Anderson, E., Sethi, S., Steinschneider, S., Flecker, A., & Gomes, C. Efficiently Approximating the Pareto Frontier: Hydropower Dam Placement in the Amazon Basin. In *Proceedings of the AAAI Conference on Artificial Intelligence*, New Orleans, LA, USA, 2–7 February 2018; Volume 32. DOI: 10.1609/aaai.v32i1.11347

Yağ, İ., & Altan, A. (2022). Artificial intelligence-based robust hybrid algorithm design and implementation for real-time detection of plant diseases in agricultural environments. *Biology (Basel)*, *11*(12), 1732. DOI: 10.3390/biology11121732 PMID: 36552243

Yanamala, A. K. Y., & Suryadevara, S. (2024). Emerging Frontiers: Data Protection Challenges and Innovations in Artificial Intelligence. *International Journal of Machine Learning Research in Cybersecurity and Artificial Intelligence*, *15*(1), 74–102.

Yang, T.-J., Chen, Y.-H., Emer, J., & Sze, V. A Method to Estimate the Energy Consumption of Deep Neural Networks. In *Proceedings of the 51st Asilomar Conference on Signals, Systems, and Computers*, Pacific Grove, CA, USA, 29 October–1 November 2017; IEEE: Piscataway, NJ, USA, 2017; pp. 1916–1920 DOI: 10.1109/ACSSC.2017.8335698

Yang, Z., Sun, J., Zhang, Y., & Wang, Y. (2017). Green, Green, It's Green: A Triad Model of Technology, Culture, and Innovation for Corporate Sustainability. *Sustainability (Basel)*, *9*(8), 1369. DOI: 10.3390/su9081369

Yang, Z., Sun, J., Zhang, Y., & Wang, Y. (2018). Peas and carrots just because they are green? Operational fit between green supply chain management and green information system. *Information Systems Frontiers*, *20*(3), 627–645. DOI: 10.1007/s10796-016-9698-y

Yang, Z., Sun, J., Zhang, Y., & Wang, Y. (2019). Perceived fit between green IS and green SCM: Does it matter? *Information & Management*, *56*(7), 103154. Advance online publication. DOI: 10.1016/j.im.2019.02.009

Yohannes, A., & Mauritsius, T. (2022). Critical Success Factors in Information Technology Projects. *International Journal of Emerging Technology and Advanced Engineering*, *12*(7), 45–67. DOI: 10.46338/ijetae0722_06

Zhang, Y., Macke, W., Cui, J., Urieli, D., & Stone, P. (2021). Learning a Robust Multiagent Driving Policy for Traffic Congestion Reduction. *Neural Computing & Applications*, •••, 1–14.

Zhang, Y., & Wei, F. (2021). SMEs' charismatic leadership, product life cycle, environmental performance, and financial performance: A mediated moderation model. *Journal of Cleaner Production*, *306*, 127147. DOI: 10.1016/j.jclepro.2021.127147

Zhao, Y., Chetty, G., & Tran, D. (2019). Deep learning with XGBoost for real estate appraisal. Proc. IEEE Symp. Ser. Comput. Intell., pp. 1396-1401. DOI: 10.1109/SSCI44817.2019.9002790

Zhou, X., Zhang, W., Chen, Z., Diao, S., & Zhang, T. (2021). Efficient Neural Network Training via Forward and Backward Propagation Sparsification. *Advances in Neural Information Processing Systems*, *34*, 15216–15229.

Zimmer, B. (2018, January). LISA: A practical zero trust architecture. In *Proceedings of Enigma*, Santa Clara, CA, USA.

Zorrilla, R. A. (2022). *Ethics and the use of algorithms and artificial intelligence by Amazon: A value analysis*. Retrieved from https://repositorio.comillas.edu/rest/bitstreams/527617/retrieve

About The Contributors

Fernando Ortiz-Rodriguez is a Full Professor and Head of the Artificial Intelligence and Innovation Lab at Tamaulipas Autonomous University. He was the Executive Director at the International Institute of Studies (IIES). He created the First Business School in Tamaulipas, Mexico. He worked as the Information Technology Manager at Emerson Electric, where he developed more than 40 pieces of software, some of them used globally in Emerson, and, achieved technology convergence by implementing the first efforts on IoT and Industry 4.0. Fernando is a member of National Systems Researchers (SNI) Level 1 of the National Council of Science and Technology (CONACYT) and INDEX IT Advisor

Sanju Mishra Tiwari has worked as a Research Associate for a sponsored research project "Intelligent Real-time Situation Awareness and Decision Support System for Indian Defense" funded by DRDO in the Department of Computer Applications, National Institute of Technology, Kurukshetra. Her current research interests include Knowledge-Based Systems, Ontologism, Intelligent Decision Support, Semantic/ Intelligent Query Search Engines, Intrusion Detection, and Swarm Intelligence. She has to date published 19 research papers and one 1 book chapter with international and national publishers. She has 10 years of experience in the teaching field and 2 years in the research field for developing ontologies. She is a member of IEEE. She is working as an organizing committee member of MIR Labs, USA. Further Current Affiliations: August 2022-Present: ORKG Curator at TIB Hannover Germany. April 2021-Present: DAAD Post-doc-net AI fellows. Feb 2021- Present: Adjunct Professor at Vardhman Engieering College, Hyderabad, India

Arturo Mora-Olivo I am a researcher interested in the ecology, taxonomy, floristics, biogeography, and conservation of vascular plants, especially aquatic plants.

Sashikala Mishra, member of ACM, ACMW, ISTE and Senior Member of IEEE, earned her Ph.D. in 2015, specializing in bioinformatics and data mining, machine learning and Artificial Intelligence Presently, she serves as a Professor in the Department of Computer Science and Engineering at Symbiosis Institute of Technology, Symbiosis International Deemed University, Pune. Her research spans diverse domains including Artificial Intelligence, Conservation Biology, Pricing Theory, Bioinformatics, Data Mining, Image Processing, and Networking. Dr. Mishra's primary focus revolves around devising and implementing algorithms for prediction, classification, and clustering. With over 63 publications in prestigious international journals across renowned publishing houses such as Springer, IEEE, Elsevier, Inderscience, Springer, and ACM. She edited the springer book Artificial Intelligence: Towards Sustainable Intelligence, 2023. Additionally, she has contributed significantly to intellectual property, having filed three patents and published two Intellectual Property Rights (IPRs). Notably, she has contributed chapters to the Biological Knowledge Discovery Handbook published by Wiley. Recognized for her outstanding contributions, Dr. Mishra has been the recipient of several accolades. Notable among them are the Prof. Indira Parikh's 50 Women in Education Leaders Award conferred at the World Education Congress in Mumbai, the Dewang Mehta Best Professor in Computer Engineering Award, and the Emerging Leader Award by the Venus International Foundation for her distinguished achievements in engineering. She has also been honored with the Distinguished Researcher Award, presented in collaboration with the Centre for Smart Modern Construction at Western Sydney University, Australia. Her current projects associated with Dr. Simi Bajaj, Sydney university, Dr. Bulent Yilmaz, GUST Google Scholar -

***

Demian Abrego-Almazán holds a Ph.D. in Administrative Sciences from the Autonomous University of Tamaulipas (UAT), has been a member of PRODEP and a certified professor in Administrative Informatics by ANFECA since 2004. Since 2017, he has been a member of the National System of Researchers (SNI). Currently, he belongs to the consolidated academic body "Information Technology and Strategy" (UAT-CA-132). As a full-time professor at the Faculty of Commerce and Administration Victoria, he teaches both undergraduate and graduate courses, actively participating in curricular reforms. During this time, he has provided tutoring and academic counseling. In 2019, he was named "Outstanding Professor" by the university in recognition of his career and achievements. Regarding scientific dissemination, he has published articles in high-impact journals, books, and peer-reviewed book chapters, focusing on information technology and management. He has also developed various research projects and technical reports, participating in national and international conferences. Additionally, he has registered 18 copyrights in the area of computer programs for the development of information systems. In human capital development, he has delivered diplomas, workshops, and conferences and has supervised undergraduate, master's, and doctoral theses. Some of these theses have received awards for their quality, and several of his students have joined the SNI. He has also collaborated with other academic bodies in research networks and served as an evaluator for various institutions such as CONACYT, PRODEP, and other research projects. Furthermore, he has participated as an external evaluator for journals, conferences, and teaching evaluation programs. He has also served as an external auditor for the electoral system of the State of Tamaulipas.

Mayra Elizabeth Brosig Rodriguez is a Professional Accountant and Auditor with a Master's Degree in Business Administration and a Doctorate in Educational Sciences. She is a researcher candidate for SNII; she is a certified professor by ANFFECA; and she has the recognition of a desirable profile by PRODEP, besides being the leader of a research group. PhD. Brosig has extensive experience in teaching and research. Her academic work ranges from published books to published articles in international and national indexed and refereed journals; she has participated in national and international congresses; she has been invited speaker in different academic events, as well as evaluator, moderator, and organizer. She is currently a full-time professor-researcher at the Universidad Autónoma de Nuevo León at the bachelor's and master's level, with projects that seek to promote education, technology, and sustainability. Her leadership and hard work have been fundamental to the advancement of knowledge in digital education.

C. Geetha is working as a Professor in the Department of Computer Science and Engineering at R.M.K. Engineering College. She is awarded Ph.D., in the field of Wireless Sensor Networks at Manonmaniam Sundaranar University,Tirunelveli, India. She is in teaching profession for about 27 years. She has presented and published number of papers in National and International Journals, Conference, Patents and Books. Her main area of interest includes Computer Networks, Sensor Networks, IoT and Artificial Intelligence

Susanta Das is a seasoned educator with a proven record of teaching and mentoring diverse students, working at multiple institutions, and providing service to the various initiatives of organizations. He received his Ph.D. and M.A. degrees from Western Michigan University (WMU), USA, and M.Sc. degree from Banaras Hindu University (BHU), India all in Physics. He continued his research as a Marie-Curie post-doctoral fellow at Stockholm University, Stockholm, Sweden on beam diagnostics for the DESIREE (Double ElectroStatic Ion Ring ExpEriment) in the project DITANET (Diagnostic Techniques for particle Accelerators – a Marie-Curie initial training NETwork), at the Indian Institute of Science Education and Research-Kolkata, India on high-pressure physics as a Project Scientist-B, and at the University of Electro-Communications, Tokyo, Japan on ion-surface interactions as a Post-doctoral fellow. Throughout his career, Dr. Das worked and collaborated with many researchers, Ph.D., master, and visiting students from different countries. He further visited many countries to discuss research and an international conference participant (UK, Italy, Germany, Greece, Belgium, Bulgaria, Romania, Brazil etc.). He co-authored several research articles in WoS/Scopus indexed international journals, conference proceedings, and scholarly book chapters published by renowned international publishing houses. Beyond his research endeavors, Dr. Das has a storied history of service to academic and administrative committees, exhibiting his commitment to the growth and development of educational institutions. His experience includes tenure at Central University South Bihar, Sri Sri University, and P.K. University before assuming his current role at Ajeenkya DY Patil University. He received the Marie-Curie post-doctoral fellowship in Sweden, Science Academies' Summer Research Fellowship in India, Gwen Frostic Doctoral Fellowship, Department Graduate Research and Creative Scholar Award by WMU, and Leo R. Parpart Doctoral Fellowship by Dept. of Physics, WMU, among many others, throughout his academic journey. At present, Dr. Das continues to delve into cutting-edge fields, with a keen interest in nanotechnology, quantum computing, and data science. His multifaceted contributions, spanning teaching, research, and administrative leadership, showcase a dedicated professional who is instrumental in advancing the vision and mission of the institutions he serves.

Gargee Gavhane, a motivated and creative student with a keen enthusiasm for geopolitics, is presently studying for a Bachelor of Technology in Robotics and Automation Engineering. A flair for interdisciplinary learning and problem-solving is evident in her academic and extra-academic activities, which she nurtures through several extracurricular activities and academic endeavors. Gargee possesses a solid grounding in physics, mathematics, and technical communication, which gives her a thorough knowledge of intricate systems and technologies. Her increasing experience in data structures and algorithm development demonstrates her commitment to learning the technical skills she needs in her profession. Apart from her scholarly pursuits, Gargee is a participant in hackathons and college projects where she uses her problem-solving abilities to tackle real-world problems. Her strategic mindset can be further seen through her presence in design competitions such as CAD Innova and high-profile competitions such as the Smart India Hackathon (SIH). Her interest in strategic thinking also extends to the game of chess, which further increases her analytical capabilities. As a problem solver and strategic thinker, Gargee is set to make invaluable contributions to her profession. With her technical background and experience dealing with intricate geopolitical matters, she is a cross-functional and progressive professional. Being well-grounded in robotics and automation, she is capable of addressing the intricacies of the world that is becoming more interdependent.

Rakesh Kumar Jain received his B.E. and M.E. in Civil Engineering from R.D. University, Jabalpur, Madhya Pradesh, and the University of Pune, Maharashtra. He earned his Ph.D. in Engineering from Rajiv Gandhi Technical University, Bhopal. He served as the Principal of Dr. D.Y. Patil Institute of Technology, Pune, India, and as the Director of Rajarshi Shahu College of Engineering, Pune, India, before joining Ajeenkya D.Y. Patil University as Vice-Chancellor. He has been associated with the National Board of Accreditation (NBA) and the National Accreditation and Assessment Council (NAAC) for several years. He received the title of "Margdarshak" (Mentor) from the All India Council for Technical Education (AICTE) for guiding institutions through the NBA accreditation process. He was also appointed as an "Ambassador" for mentoring five institutions for NAAC accreditation under the "Paramarsha" (Advisor) scheme sanctioned by the University Grants Commission of India (UGC). Additionally, he was appointed as a "Mentor" by the Directorate of Technical Education, Mumbai, India, to guide institutions for NAAC/NBA accreditation. He has received several prestigious awards, including the Hon. Balasaheb Wagh Memorial "Best Institute Award" by the Association of Management of Unaided Colleges (Maharashtra), the "Best Principal" and "Best College Award" by the Indian Society of Technical Education, New Delhi, the "Best Principal Award" by Savitribai Phule Pune University, Pune, and the "Best Affiliated College Award" by Savitribai Phule Pune University, Pune. Additionally, he was honored with the "Best Teacher and Certificate of Excellence for Leadership" in September 2014 by Dr. D.Y. Patil Vidya Pratishthan Society, Pune, and the Visvesvaraya Construction Excellence Award by Construction & Business Record, Mumbai, India. He has published books with reputed national and international publishers, including Routledge (Taylor & Francis Group, USA) and CRC Press, Boca Raton, USA. He has published or presented more than 144 technical papers on topics related to Geotechnical Engineering and Transportation Engineering in international and national journals, seminars, and conferences. Three of his papers were adjudged as "Best Paper". He holds several patents and has guided numerous doctoral students. He is a member of various professional bodies, including the American Society of Civil Engineers, the Institution of Engineers (India), and the Indian Society for Technical Education, among others. He was nominated by the Chancellor (Governor of Maharashtra) as a Member of the Academic Council of

SPPU, Pune. He has served as a Member of the Board of Studies (BOS) in Civil Engineering under the Faculty of Engineering (FOE), Chairman of BOS in Civil Engineering under FOE, and a Member of the Research & Recognition Committee. Additionally, he has been an Academic Council Member of JSPM University, Pune; Sanjivani College of Engineering, Kopargaon; M.M. College of Engineering, Pune; Sri Venkateswara COE, Tirupati; Shri Ramswaroop University, Lucknow; and the Confederation of Indian Industries (CII) Western Region Committee on Education, Planning & Monitoring Board. He was also part of the Organizing Committee for the Indo-US Conclave and the Internal Quality Assurance Cell (IQAC) of Dr. D.Y. Patil University, Pimpri, Pune. He has visited numerous countries as a delegate, conference participant, and researcher. He has also organized several Faculty Development Programs (FDPs), Short-Term Training Programs (STTPs), Summer and Winter Schools, seminars, and conferences.

Vishal Jain is presently working as Professor at Department of Computer Science and Engineering, School of Engineering and Technology, Sharda University, Greater Noida, U. P. India. Before that, he has worked for several years as an Associate Professor at Bharati Vidyapeeth's Institute of Computer Applications and Management (BVICAM), New Delhi. He has more than 14 years of experience in the academics. He obtained Ph.D (CSE), M.Tech (CSE), MBA (HR), MCA, MCP and CCNA. He has authored more than 90 research papers in reputed conferences and journals, including Web of Science and Scopus. He has authored and edited more than 30 books with various reputed publishers, including Elsevier, Springer, Apple Academic Press, CRC, Taylor and Francis Group, Scrivener, Wiley, Emerald, NOVA Science and IGI-Global. His research areas include information retrieval, semantic web, ontology engineering, data mining, ad hoc networks, and sensor networks. He received a Young Active Member Award for the year 2012–13 from the Computer Society of India, Best Faculty Award for the year 2017 and Best Researcher Award for the year 2019 from BVICAM, New Delhi.

Sujatha K is Currently working as an Assistant Professor in the Department of Computer Science and Engineering, SRM Institute of Tehnology and Science, Ramapuram, India.Se completed her B.E in Anjalai Ammal Mahalingam Engineering College, M.E in Periyar Maniammai College of Technology For Women and PhD in Hindustan Institute of Technology and Science,Chennai. She has published 4 journal papers and 7 patents. Her research interest includes security, cloud computing and fog computing.

Roop Kamal is currently working as an Assistant Professor in University Business School,Chandigarh University,Mohali,Punjab.She is a post graduate

(M.com) from Panjab University,Chandigarh.She has also qualified UGC NET in Commerce and is currently pursuing PhD in Commerce from RIMT University,Mandi Gobindgarh. She has over 7 years of experience in academics. Her research interests include Green marketing, Digital Marketing, Sustainability and Consumer Buying behaviour. She has published several research papers and articles in reputed international and national journals and has presented papers in various national and international conferences.

Archan Mitra is an Assistant Professor at School of Media Studies (SOMS) at Presidency University, Bangalore. He is the author of two book “Cases for Classroom Media and Entertainment Business” and “Multiverse and Media”, he also has other several edited books to his credit. He has done his doctorate from Visva-Bharati Santiniketan, West Bengal in the field of “environmental informatics and communication for sustainability”. In addition to that he is a certified Science Communicator and Journalism from Indian Science Communication Society (ISCOS), certified Corporate Trainer with Amity Institute of Training and Development, Certified Social Media Network Analyst. He has a strong interest in environmental communication. He was awarded certificate of merit by PRSI, Kolkata Chapter and Medal of Honor by Journalistic Club of Kolkata. He was working as a research assistant with the World Bank's “Environmental Capacity Building in Southeast Asia” project at IIM Kashipur. He was instrumental in launching the World Bank's Green MBA MOOC, he has also assisted in the research project on Uttarakhand disaster mitigation by ICSSR, the leading research on Uttarakhand disaster.

Nishakant Ojha is a distinguished academic and expert in emerging technologies, management, and interdisciplinary studies. Below is a summary of his expertise and qualifications: Academic Expertise 1. Cybersecurity: Renowned for his deep knowledge of cyberspace, cyber defense mechanisms, and strategies for national and organizational security. 2. Artificial Intelligence (AI): Specialized in AI-driven technologies, focusing on their application in governance, defense, and strategic industries. 3. IoT and Emerging Technologies: Postdoctoral research focused on AI, IoT, and new technologies, emphasizing their integration into transformative solutions. 4. Quantum Technologies: Expertise in the principles and applications of quantum computing and quantum communication, especially as they relate to national security and technological advancement. 5. Aerospace Studies: His academic contributions in aerospace technologies, coupled with his role as a visiting professor at the National Aviation University in Kyiv, underline his authority in this domain. 6. Management: Holds a doctorate in Management, integrating innovative strategies with AI and leadership for organizational advancement. Qualifications

1. Postdoctoral Research: Completed postdoc at Aalborg University, specializing in AI, IoT, and emerging new technologies. 2. Doctorate: Holds a doctorate in Management, demonstrating expertise in leadership and organizational strategy. 3. Law Degree: Holds a law degree with a specialization in cyber law, bridging technology and legal frameworks. 4. Ph.D. Supervision: Successfully supervised 4 PhD students, primarily in the areas of Artificial Intelligence (AI) and Management. 5. Visiting Professorship: Served as a visiting professor at the National Aviation University, Kyiv, Ukraine, enhancing his academic credentials in aerospace studies. 6. Leadership Roles: Served in top management and policy-level positions in the government, highlighting his expertise in strategic decision-making and governance. Dr. Ojha's extensive academic and professional achievements demonstrate his remarkable contributions to research, education, policymaking, and leadership across multiple disciplines.

Nithi Ravya S is a third-year B.Tech student specializing in Artificial Intelligence and Data Science. She is passionate about leveraging AI and emerging technologies, dedicating herself to exploring innovative solutions in data-driven decision-making and intelligent automation. With a keen interest in machine learning, IoT, and the evolving tech landscape, she aspires to make impactful contributions to the field of AI.

Lalitha S D, I am working as an Associate Professor in CSE department at R.M.K. Engineering College, Kavaraipettai. I have completed my Ph.D. in Anna University, Chennai by November 2023. I have around 22 years of teaching experience. My area of interests are Web Programming & Artificial Intelligence.

Maria Ines Salas Rubio, Bachelor's degree in Administrative Computing, Master's degree in Information Systems and PhD in Strategic Business Management from the Autonomous University of Tamaulipas. In the private sector, she has worked in the sales area, data entry and software development in various organizations, she has served as an entrepreneur running a computer rental business with hardware and software maintenance service from 2001 to 2006 and as a micro-entrepreneur of women's items from 2008 to 2018. Her start in teaching was in 1994 in the private sector as a software trainer for the business area and at the University from 2000 participating in academic bodies in student projects, she worked in the Business Incubator of the Faculty of Commerce and Administration of Tampico (FCAT) from 2004 to 2008 as coordinator of entrepreneur training, she served as a collaborator in the Department of Coordination of National and International Accreditations from 2018 to 2019. She has participated as a speaker in national and international conferences and in events such as Mentoring Walk Tampico and Researcher's

Day. In recent years she has collaborated on research projects and published with professors from the Autonomous University of Tamaulipas, the National Autonomous University of Mexico, the Autonomous University of San Luis Potosí and the Autonomous University of Nuevo León; the latter being where she also did a research stay. She is currently a Research Professor at FCAT, teaching classes to the 3 undergraduate degrees (LA, LNI and CP) and postgraduate classes (MGEE). She represents the Women-UAT liaison in her faculty and is part of the academic body of Management of Intangibles in Organizations. Her lines of research are the use of ICT, electronic commerce and consumer behavior. She is certified by the National Council for Standardization and Certification of Labor Competencies (CONOCER) in the Design of Didactic Strategies applying ICT 2023-2026. Finally, Dr. Salas has obtained the distinction as a member of the National System of Researchers as a candidate for SNI 2022-2025.

Atharva Saraf is pursuing his undergraduate studies in Computer Science at the School of Engineering, Ajeenkya DY Patil University. His research interests are cybersecurity, artificial intelligence, and data science. He has published book chapters in reputed national/international publishers such as CRC Press and IGI Global, and presented his research results at various conferences.

Kamaljeet Kaur Siddhu has done Ph.D in Marketing. She is a Fellow Member of the Institute of Company Secretaries of India. She is Alumni of IIM Kashipur. Basically, a Master's in commerce, she is also a Law graduate and has done several PG diploma and certification courses including PG. in Psychology, New Labor Codes, certified CSR & ESG Professional, ISO 9001:2015 Lead Auditing, etc. She has more than 20 years' experience in the field of Corporate Affairs, Legal & HR. She is currently working as Executive Vice President of Ajeenkya D Y Patil Group and earlier worked as Director – Technical Campus at D.Y. Patil Knowledge City, Lohegaon. Few of her previous assignments were as Director- Corporate Affairs with a Private University, Vice- President – Corporate Affairs, Legal & HR with City Corporation Limited (Makers of Amanora Park Town); Legal Head & Company Secretary with Manufacturing Construction, and IT Companies. She is also a Member of Arbitration, Mediation & Conciliation Network and has assisted Dr. K. R. Chandratre, past president of ICSI in writing a book on "Sections- 397 & 398 of the Companies Act- Oppression and Mismanagement and "Cases and Material on Company Law. She is a Division Director in D-125 of Toastmasters, two times Triple Crown Award Winner and Division level International Speech Contest winner in District 98 of Toastmasters. She is also engaged as Visiting Faculty for teaching Management subjects to MBA Students; and has done paper presentations in various National Seminars and has acted as External Examiner for various Management Institutes. She has served as President of Rotary Club of Pune Hadapsar (2018-19) and works actively for Rotary District 3131 doing social work in the areas of Literacy, Environment and Women Empowerment. She is currently Co-Director for Peace Building & Internal Conflict in Rotary District 3131. She is also a freelance career and psychological counsellor. Her other skills include Counselling, Behavior Therapy, Competency Mapping, Transactional analysis & Neuro Linguistic Programming, Institutional and Corporate Trainings, Legal Advisory, POCSO awareness trainings, Management Consulting, POSH awareness trainings, Drafting of Company Manuals, Risk Management, Corporate Governance, CSR Policy, Labor Law Compliances; Profitability Management, QMS, EMS, HSE awareness training, train the trainer and other tailormade training programs for individuals as well as organizations. She has conducted several workshops and coahing programs such as Teachers training Program as per new Education Policy, "Women Empowerment", Invincible Leader Program, Leaderships skills, Negotiation skills, Transactional Analysis, Motivation, Diversity & Inclusion, Customer Handling Techniques, Conflict Management, Etiquettes, Empathy, Conflict Resolution, Goal setting, Law of attraction, Neuro Linguistic Programming, Cognitive behavior therapy, Career Guidance, Time Management, 7 habits of highly effective people, Selling Skills, POSH, POCSO, Stress Management, Brand Management, Communication Skills, Anger Management, Motivation Essential Life skills and tailor-made trainings for Individuals, Institution as well as Corporates.

Alishba Siddiqui is currently pursuing her B.Tech. in Computer Science at the School of Engineering, Ajeenkya DY Patil University, Pune, India. Her research interests are UNSDGs, AI, cybersecurity, and data analysis.

Rebika Singh is a seasoned Human Resources Business Partner & HR Leader with over 11 years of experience in strategic HR management, change management, and organizational development. She has successfully led HR transformations across industries such as education, sports, film & media, agriculture, defense research & innovation, automobile, electric vehicles (EV), engineering, IT, construction, and steel. With a deep understanding of people management, talent acquisition, policy formulation, performance management, and training & development, Rebika has played a pivotal role in shaping HR functions in both large multinational corporations and fast-growing startups. Rebika Singh holds a Postgraduate degree in HR and Marketing, following her Bachelor of Computer Applications (BCA) from BIT Mesra. Her strong academic foundation, combined with extensive hands-on experience, has enabled her to excel in business transformation through HR strategies. Her career spans multiple locations across India, including Bhubaneswar, Delhi, Pune, and Ahmedabad, where she has been instrumental in driving HR excellence and workforce development. As the HR Head at Ajeenkya D Y Patil Group, a leading business conglomerate with diversified interests, she played a strategic role in: Driving change management initiatives, ensuring smooth transitions and HR transformation across various domains such as education, sports, film & media, agriculture, and defense research & innovation, Leading organizational development (OD) programs, enhancing workforce productivity and business efficiency, Developing HR strategies tailored to diverse industries, ensuring business-specific talent solutions for long-term growth, Implementing HR frameworks that fostered leadership development, employee engagement, and process optimization, to name a few.

Index

www.ingramcontent.com/pod-product-compliance
Lightning Source LLC
LaVergne TN
LVHW061542290926
849162LV00004B/885

* 9 7 9 8 3 6 9 3 8 6 4 3 9 *